AQUINAS AND THE SHIP OF THESEUS

AQUINAS AND THE SHIP OF THESEUS

SOLVING PUZZLES ABOUT MATERIAL OBJECTS

CHRISTOPHER M. BROWN

continuum
LONDON • NEW YORK

Continuum
The Tower Building, 11 York Road, London SE1 7NX
15 East 26th Street, New York, NY 10010

British Library Cataloguing-in-Publication Data
A catalogue record for this book is available from the British Library.

ISBN: HB: 0-8264-7828-X

Library of Congress Cataloging-in-Publication Data
A catalog record is available from the Library of Congress.

Typeset by Aarontype Limited, Easton, Bristol
Printed and bound in Great Britain by MPG Books Ltd, Bodmin, Cornwall

CONTENTS

PREFACE

In the last twenty-five years a number of philosophers have argued that classical puzzles about material objects such as the Ship of Theseus show that our common-sense intuitions about compound material objects are *logically incompatible* with one another. It seems to these philosophers that we have to either give up one of our common beliefs about material objects or else give up reasoning about material objects altogether. Although there are a number of solutions to Ship of Theseus-style puzzles that have garnered support among contemporary analytic philosophers, virtually all of these solutions argue for the rejection of a common-sense intuition about material objects. Whatever else may be said therefore in defence of these particular solutions, they all entail revising our common-sense understanding of a material object.

One way of approaching the project of solving puzzles about material objects involves first re-examining some of our own fairly entrenched metaphysical assumptions about such objects. This re-examination project is itself ably facilitated by comparing contemporary views on material objects with those of a great metaphysician from the past, someone who may not share all of our contemporary philosophical predilections. Indeed, some of Thomas Aquinas's views on the composition of material substances are at variance with metaphysical assumptions commonly presupposed by contemporary analytic philosophers and Aquinas's views suggest some interestingly novel ways of handling classical and contemporary puzzles about material objects.

In this work I propose that Aquinas's metaphysic of material objects (or what is sometimes referred to as his 'natural philosophy') shows that it is possible to solve Ship of Theseus-style puzzles and still hold on to all of our common-sense intuitions about material objects. Although Aquinas never explicitly tackles problems such as the Ship of Theseus puzzle – probably because Aristotle does not have much to say about such matters – what Aquinas does have to say about substances, composition and identity allows one to construct a set of Thomistic solutions to the famous puzzles about material objects. These Thomistic solutions do not entail the falsity of any of our common-sense intuitions about material objects.

I argue here in several stages for the relevancy of Aquinas's thought for the solving of puzzles about material objects. The first two chapters show how contemporary analytic philosophers approach philosophical problems having to do with the nature of material objects. More specifically, in the first chapter I both explain how certain common-sense intuitions about compound material objects are supposed to lead to problems in the first place and I offer a survey of the most important contemporary solutions to those problems. The second chapter takes a detailed look at some arguments about material objects put forward by three important contemporary metaphysicians: Peter van Inwagen, Lynne Rudder Baker and Dean Zimmerman. Having clarified how contemporary analytic philosophers approach the classical puzzles about material objects, I offer my own exegesis of Aquinas's metaphysic of material objects in the next several chapters. The third chapter introduces Aquinas's metaphysic of material objects through an examination of a concept crucial to his philosophical enterprise, the concept of material substance. The fourth chapter discusses Aquinas's views on material composition and part-hood. The fifth chapter explains Aquinas's views on the identity and indivi-duation of material objects through time and change. The final two chapters of the work bring my exegesis of Aquinas's metaphysic of material objects to bear on puzzles about material objects. In particular, the sixth chapter shows how Aquinas's views on material composition and identity allow him to solve classical and contemporary puzzles about material objects in ways that pre-serve our common-sense intuitions about material objects. Finally, the seventh chapter defends these Thomistic solutions to puzzles about material objects against some possible objections.

I would like to thank those people who have had a hand, in one way or another, in helping me to complete this book on Aquinas's metaphysic of mate-rial objects. Thanks to one of my first philosopher teachers, Arvin Vos, who first sparked my interest in Aquinas's thought. The following people provided help-ful comments either on portions of the manuscript or on papers that are ances-tors to it: Adiel Brasov, Jeff Brower, Merry Brown, Chris Calloway, John Claiborne, Bryan Cross, Darin Davis, William Eaton, Jason Eberl, Miguel Endara, James Fieser, Alicia Finch, Norman Lillegard, Scott MacDonald, Steven Napier, Robert Pasnau, William Rehg, Karl Schudt, Kevin Timpe, Joshua Thurow and Christina Van Dyke. Father Theodore Vitali and Colleen McCluskey read and commented upon an early version of the entire manu-script. I offer special thanks to Eleonore Stump. She not only gave me extensive comments on a number of early drafts of the entire manuscript but has also been a great mentor to me, both professionally and scholastically. I would like to thank the entire department of Philosophy at Saint Louis University for pro-viding such an intellectually fertile and caring place in which to receive a graduate philosophical education. Also, several grants that I received while a student at Saint Louis University made it possible for me to work on papers (and a dissertation) that are ancestors to this book, namely, a 2000–2001 SLU 2000 fellowship and a 2001–2002 dissertation fellowship. Thanks also to my colleagues in the department of History and Philosophy at the University of Tennessee at Martin for assistance and encouragement during the time that I was working on the final revisions of the manuscript. In this regard I would especially like to thank Jim Fieser, Norman Lillegard, David Coffey and Donna Cooper Graves.

I reserve the greatest thanks for God and my family. For all of her support and love through the years, I thank my wife Merry Elizabeth. Merry is one of the most multi-faceted persons I know, a wonderful wife, mother, teacher – and philosopher. I couldn't have done any of what I attempt to do in this work without her patient love and support. Thank you to my parents, Noel and Patricia Brown, my parents-in-law Ed and Susan Hill, and my siblings Noelle and Brandon for always being there. Finally, I would like to thank my sons Judah Christopher and Leopold Edward simply for being. I dedicate this book to them.

LIST OF ABBREVIATIONS

CT	*Compendium theologiae*
DEE	*De ente et essentia*
DME	*De mixtione elementorum*
DPN	*De principii naturae*
In BDT	*Expositio super librum Boethii De Trinitate*
In DA	*In Aristotelis librum De Anima commentarium*
In DGC	*In librum primum De generatione et corruptione expositio*
In Met.	*In duodecim libros Metaphysicorum Aristotelis expositio*
In Phys.	*In octo libros de physico auditu sive Physicorum Aristotelis commentaria*
In Sent.	*Scriptum super libros Sententiarum*
QDA	*Quaestio disputata de anima*
QDP	*Quaestiones disputatae de potentia*
QDSC	*Quaestio disputata de spiritualibus creaturis*
QDUVI	*Quaestio disputata de unione verbi incarnati*
QQ	*Quaestiones quodlibetales*
SCG	*Summa contra gentiles*
ST	*Summa theologiae*

1

The Problem of Material Constitution

Imagine a ship, whose sole function is to make a yearly voyage to a neighbouring country in order to honour a heroic deed from the past. The ship in question is composed of wooden planks, and her shape might be described as very distinctive. After a few years of making her yearly voyage, the ship's planks begin to weather. The crew decides that henceforward, before the ship sets sail each year, they will replace the weathered planks of the ship with new ones. Eventually, all of the planks of the original ship are replaced. Now someone (say her name is Merry) collects the planks that are disposed of from the original ship each year, until some years later, Merry has collected all of the planks from the original ship. Furthermore, Merry decides to put the planks she has collected together in her back yard, giving those planks the same distinctive configuration they had when they composed the original ship at the time of her first voyages. Given the information in this story, someone might well wonder which ship is numerically identical to the *original ship*. Is it the *continuous ship*, which continues to make the yearly voyage to the neighbouring country and whose spatio-temporal history is continuous with that of the original ship, or is it the *reconstructed ship*, which is composed of the same set of planks as the original ship? Indeed, they cannot both be numerically *identical* to the original ship, since the continuous ship is out to sea, and the reconstructed ship resides in Merry's (very dry) back yard!

The story is, of course, a re-telling of the famous Ship of Theseus puzzle. Originally discussed among Greek philosophers as a puzzle that raises questions about the possibility of an object surviving the replacement of its parts,[1] Thomas Hobbes later records a version of the puzzle that includes mention of the reconstructed ship. In Hobbes's formulation of the Ship of Theseus puzzle the question now becomes, which ship – the continuous or the reconstructed – is numerically identical to the original ship?[2]

In the last twenty-five years a number of philosophers have argued that classical puzzles about material objects such as the Ship of Theseus show that our common-sense intuitions about compound material objects are *logically incompatible* with one another.[3] These common-sense intuitions about material objects ('IMO' for short) include the following:

(IMO1)   There are such things as *compound material objects*, that is, there are material objects that are composed or constituted of other material objects.

(IMO2) There are many, many different kinds of compound material objects, including different kinds of atoms, molecules, aggregates of atoms or molecules, proteins, enzymes, plants, animals, tissues, organs, limbs, body sections, artefacts and artefact parts.

(IMO3) Compound material objects endure through time and change.

(IMO4) There are compound material objects that can survive certain losses, gains and replacements with respect to their parts.

(IMO5) Two material objects cannot exist in the same place at the same time.

(IMO6) Necessarily, identity is a transitive relation.

Many contemporary philosophers think that these common-sense intuitions are logically inconsistent – at least when they are combined with certain other apparently plausible theses about material objects. Following Michael Rea, I refer to this apparent conflict between common sense and logic where the nature of material objects is concerned as 'the Problem of Material Constitution' (PMC).[4] Solving the PMC involves either offering good reasons for giving up one rather than another intuition about material objects or else showing that the PMC is only an apparent problem.

Common-sense intuitions about compound material objects

Before explaining in detail how the intuitions listed above are supposed to be logically inconsistent, I want to first say something more about what I have in mind by each of these intuitions. All compound material objects are thought, intuitively, to have certain features. Some of the more important of these features are captured in IMO3, IMO4 and IMO5.

For example, IMO3 and IMO5 stress features that are thought to be shared by all material objects. IMO3 has it that material objects can remain identical through time and change. That is to say, it is possible – in fact it actually occurs all of the time – that a material object exists as a whole at two different times, even though that material object as a whole has different properties at those two times. It would be counter-intuitive to say that only a part of Socrates exists at time t. Rather, Socrates as a whole exists at t and at $t + 1$, although Socrates has different physical properties at those two times, e.g. Socrates is composed of different particles at those two times.

IMO5 captures the intuition that a *material* object is the sort of object that excludes any other object from existing in the very same place (at the same time) that it does. My refrigerator and my couch can't co-occupy the same place at the same time. Similarly, the arguments of some contemporary philosophers notwithstanding, a statue and a piece of clay – if they are both *material* objects – cannot co-occupy the same place at the same time. The impossibility of two objects being spatially coincident is built into our common-sense notion of what it means to be something material.

In contrast to IMO3 and IMO5, IMO4 expresses an intuition specifically about *compound* material objects: many (but perhaps not all) compound material objects are able to survive the loss, addition or replacement of a part. Take a prime example of a compound material object such as a tree. One of the features of a tree that is commonly thought to make it something different from, say, an aggregate of rocks, is that the tree, and not the aggregate, can survive the loss of one of its parts. The tree is composed of one set of atoms at one time, but at a future time, the numerically identical tree is composed of a different set of atoms. In fact, we think that an organism that completely changes out the atoms that compose it might still be the numerically same organism.

IMO6 is not an intuition about material objects *per se*; rather, it is a logical intuition, and it is one that figures prominently in puzzles about material objects. Here what is at issue is the extent to which the transitivity of identity holds true. I claim that the necessity of the transitivity of identity is part of the canon of common sense. If x is identical to y, and y is identical to z, then it would be counter-intuitive to say that x might not be identical to z.

Of course, common sense also has it that that there are such things as compound material objects (IMO1). But common sense also has it that there are many, many different *kinds* of compound material objects (IMO2). Following Lynne Rudder Baker, I will label the view that accepts IMO2, 'ontological pluralism'.[5] As we shall see, philosophical commitment to ontological pluralism is subject to more or less conservative versions. But IMO2 intentionally captures views central to all ontological pluralists: there are not only *living* compound material objects, but also non-living compound material objects, some of which are naturally occurring, while others are artificial. Furthermore, there exist various kinds of parts of both natural and artificial compound material objects – and these parts are themselves kinds of compound material object, e.g. hands, feet, walls and roofs. The advocate of IMO2 (or ontological pluralism) not only accepts that the compound material objects studied by the special sciences exist, but she also accepts the existence of the compound material objects of our everyday experience.

The arguments and positions put forward by contemporary philosophers in discussing the PMC are, more or less, intended to save as many of these common beliefs about material objects as possible. However, as we shall see, most contemporary solutions to the PMC involve conceding that at least a piece of our common-sense picture of material reality be rejected.

The Ship of Theseus and the PMC

How exactly are the above-mentioned intuitions about material objects supposed to form an inconsistent set? Consider the most famous of the classical puzzles about material objects, the Ship of Theseus puzzle.[6] It can be formulated in such a way so as to show that, when common-sense intuitions about material objects are combined with an apparently uncontroversial notion of 'aggregate', a contradiction is the result. The Ship of Theseus puzzle therefore can be formulated such that it explicitly raises the PMC (call this formulation of the Ship of Theseus puzzle, 'SOT').[7] SOT employs the following nomenclature. At time t, we have what I have referred to as 'the original ship'. Now the original ship is composed of an aggregate of planks at t. Call this aggregate, 'aggregate$_{OS}$'. At time $t+1$, we have what I have referred to as 'the continuous ship' and 'the reconstructed ship'. These ships are also each composed of an aggregate of planks (call them 'aggregate$_{CS}$' and 'aggregate$_{RS}$', respectively). SOT thus runs as follows:

(1)　The original ship at t is numerically identical to aggregate$_{OS}$ [from IMO3 and IMO5].

(2)　Aggregate$_{OS}$ is numerically identical to aggregate$_{RS}$ [from the assumption that aggregates x and y are numerically identical if, and only if, x and y have all and only the same proper parts].

(3)　Aggregate$_{RS}$ is numerically identical to the reconstructed ship at $t+1$ [from IMO3 and IMO5].

(4)　Therefore, the original ship at t is numerically identical to the reconstructed ship at $t+1$ [from 1–3 and IMO6].

(5)　The original ship at t is numerically identical to the continuous ship at $t+1$ [from the fact that the original ship at t and the continuous ship at $t+1$ are spatio-temporally continuous and common-sense intuitions IMO2, IMO3 and IMO4].

(6)　Therefore, the reconstructed ship at $t+1$ is numerically identical to the continuous ship at $t+1$ [from 4 and 5 and IMO6].

(7)　The reconstructed ship at $t+1$ is not numerically identical to the continuous ship at $t+1$ [from the impossibility of bi-location].

Since 6 and 7 are contradictory opposites, 6 follows logically from 1–5, and 7 looks undeniably true – it is impossible for a material object to bi-locate – we must reject one of the other premises of SOT (on pain of contradicting ourselves).

But premises 1–5 of SOT – or at least beliefs which provide the basis for accepting such premises – look as though they are true as a matter of common sense. First, there *are* such things as artefacts and their parts (IMO1 and

IMO2). Second, if, for example, the original ship and aggregate$_{OS}$ are *not* identical, then it is possible for two material objects to exist in the same place at the same time. But it is not possible for two material objects to exist in the same place at the same time (IMO5).[8] So premises 1 and 3 are true. Third, compound material objects such as ships endure through the replacement of their parts (IMO3 and IMO4); premise 5 is true. Finally, identity is necessarily a transitive relation (IMO6) and so 4 and 6 are safe inferences. Hence, by assuming the apparently uncontroversial views that a material object cannot be bi-located and that aggregates x and y are numerically the same if and only if x and y have the same proper parts,[9] common-sense assumptions IMO1–IMO6 appear to land us in a contradiction.

The Ship of Theseus puzzle – and other puzzles about material objects I'll have occasion to mention in the sequel – can thus be formulated to show that the common-sense assumptions we typically make about the compound material objects of our everyday experience apparently lead us into contradictions. This is the Problem of Material Constitution, or PMC, for short.

Solving the PMC: a glance at the contemporary approaches

Given that there are a certain number of assumptions that together generate the PMC, there will therefore be a corresponding number of different ways of solving the PMC. Indeed, there are at least six – the denial of any of the six common-sense intuitions I have been speaking about, along with a defence of such a denial, constitutes an attempted philosophical solution to the PMC. Rather than speaking about every possible way of solving the PMC here, I take a look at those solutions to the PMC that have garnered the most attention from contemporary philosophers. As we shall see, these solutions always involve offering an explanation for why it is reasonable to *deny* a common-sense intuition about compound material objects.

The *standard account* is a theory of material constitution that admits the possibility of spatially coincident objects.[10] To say that spatially coincident objects are possible is to say that two individual material objects can exist in the very same place at the very same time. The advocate of the standard account thus solves any puzzle that raises the PMC by rejecting IMO5.

As Michael Burke has pointed out, the standard account is the most common way of handling the PMC among contemporary philosophers.[11] Why do so many analytic philosophers find this way of thinking about material constitution so compelling? Perhaps it can be said that the standard account solves the PMC by doing the least amount of damage to our common-sense view of material reality. For example, it is compatible with a robust ontological pluralism, the notion that material objects exist as material wholes at every time in which they exist, and the idea that identity is necessarily a transitive relation.

But certainly the belief 'one material object to a place' is also a part of the canon of common sense. Why abandon this belief? It might be suggested that contemporary physical theories offer reasons for accepting the possibility of spatially coincident material objects. Contemporary physicists sometimes talk as though fundamental particles, e.g. quarks and electrons, have the ability to be spatially coincident. Thus, this gives one reason to think it is at least *possible* for two material objects to wholly co-occupy the same space at the same time. If it is possible for sub-atomic particles to be co-located, why not macro-level material objects? Of course, this argument assumes – controversially – that fundamental particles such as quarks and electrons are individual material objects. Furthermore, even if quarks and electrons are usefully described as individual material objects, it may be that what holds true for them doesn't hold true for macro-level material objects, such as tables, trees and tigers.

The most important philosophical argument supporting the standard account takes its start from the observed *persistence conditions* of ordinary material objects. An object's persistence conditions are those properties of an object that pick out the changes that it can and can't undergo and still remain in existence. A statue and the clay that constitutes it appear to have different persistence conditions, e.g. a statue can survive losing one of its parts – think the Venus de Milo – but a piece of clay can't survive the loss of one of its parts and remain the numerically same piece of clay. However, an object's persistence conditions are included in that object's *essential properties*, namely, those properties that an object has in every possible world in which it exists. So a statue and the piece of clay that constitutes it have different essential properties in virtue of having different persistence conditions. But objects having different essential properties are non-identical. Thus, a statue and the piece of clay that constitutes it are non-identical compound material objects. This despite the fact that they are spatially coincident objects – that is, a statue and the piece of clay that constitutes it are both composed of the same set of fundamental particles.

Contrary to what every puzzle that raises the PMC would suggest, the advocate of the standard account thinks that spatially coincident objects are indeed possible. For example, the Ship of Theseus (a kind of object that can survive the loss of a part) at any time in which it exists is constituted by an aggregate of planks (a kind of object that can't survive the loss of one of its parts). The advocate of the standard account rejects premises 1 and 3 of SOT because they assume that the relationship between a ship and the aggregate of planks that constitutes it is identity. Recall that denying that the ship and the aggregate of planks are identical suggests that there are therefore two non-identical objects existing in the same place at the same time wherever there is a ship. This amounts to a denial of IMO5. The advocate of the standard account is thus willing to reject the common-sense idea 'one material object to a place'.

Burke has argued that the standard account suffers from at least three problems. First, it unnecessarily multiplies entities. As Burke puts it, the standard account violates the principle of parsimony.[12] David Lewis sums up this criticism nicely when he says, 'It reeks of double-counting to say that here we have a dishpan, and we also have a dishpan-shaped bit of plastic that is just where the dishpan is . . . .'[13] Of course, this criticism ignores what the advocate of the standard account takes to be obvious: cases of material constitution involve a relation between material objects that have different essential properties. Thus, this criticism does not stand on its own. The critic bringing this argument against the standard account owes us an explanation as to how the statue and the piece of clay that constitutes it could be identical given that statues and pieces of clay have different persistence conditions (and thus different essential properties).

Second, the standard account 'is at odds with the commonsensical principle of one thing to a place'.[14] Here the concern is simply with the standard account's espousal of coincident objects. Someone might remark that there is nothing strange about two objects *differing in kind* occupying the same region of space at the same time; coincident objects are only objectionable when the objects in question belong *to the same kind*.[15] However, there is certainly something strange about objects of differing kinds being spatially coincident. Try to imagine a couch and a refrigerator being coincident. Doesn't this repugnance hold for all (at least macro-level) material objects? I have suggested that the answer to this question is 'yes'; the common-sense belief that coincident objects are not to be countenanced is expressed in IMO5.

Finally, Burke argues that cases of material constitution simply don't warrant the belief that there are two objects existing in the same place at the same time, given the concession on the part of the advocate of the standard account that the constituted object and the constituting object are composed of the very same parts. According to Burke, a statue and the piece of clay that constitutes it 'are qualitatively identical. Indeed, they consist of the very same atoms. What, then, could *make* them different in sort?'[16]

Among those who have responded to Burke's arguments, Lynne Rudder Baker has proposed that the difference in sort between a statue and the piece of clay that constitutes the statue can be explained in virtue of the fact that these objects have different essential properties, and an object's having the essential properties that it does is a primitive truth about that object.[17] Furthermore, pieces of clay and statues have different essential properties because statues – but not pieces of clay – derive their existence from the

intentions of rational agents. This is why two objects that share all of their physical properties in common don't necessarily share all of their qualitative properties as Burke suggest.

The standard account of material constitution solves puzzles about material
objects that raise the PMC by denying some of the crucial identity claims in
those puzzles. The Ship of Theseus is not identical to the aggregate of planks
that constitutes it at some time, since ships and aggregates have different essential properties. The high price paid by the standard account is the denial of
IMO5: since the Ship of Theseus and the aggregate of planks that constitutes
it are not identical, there are always two objects – and not one as we would
have thought – existing wherever the Ship of Theseus currently resides.

The *temporal-parts* or *four-dimensionalist account* of material objects provides
another way of solving the PMC that many contemporary philosophers find
attractive. According to this view, material objects have temporal parts in
addition to their spatial parts. Traditionally, material objects have not been
supposed to have temporal parts. Rather, we might say that common sense has
it that material objects exist in three dimensions while they move through the
fourth dimension of time. To put it another way, material objects, as long as
they exist, *endure* through time. For example, what we believe about houses
implies that it is possible that a whole house h exists at t, a whole house $h1$
exists at $t+1$, h and $h1$ have different properties (e.g. h is painted white, while
$h1$ is painted pink), and h and $h1$ are numerically identical. 'h' and '$h1$' don't
refer to different parts of one and the same house. Rather, 'h' and '$h1$' both
refer to one and the same complete house, albeit at two different times. This
common-sense way of thinking about the way material objects are related to
time and change is expressed in IMO3.

The four-dimensionalist or temporal-parts theorist rather believes that a
material object such as a house exists as a whole only over the whole temporal
course of its existence. So to take my example above of a house painted white at
one time and pink at another, 'h' and '$h1$' actually refer to two different temporal stages or parts of one house for the temporal-parts theorist. Neither h nor
$h1$ is identical to a house for the temporal-parts theorist; instead, the house of
which h and $h1$ are temporal parts is identical to $h + h1 +$ whatever other temporal parts the house in question happens to have. Thus, rather than speaking
of material objects 'enduring' though time and change (which suggests that a
whole material object is passing through time and gaining and losing properties as it goes), the temporal-parts theorist says that objects 'perdure' through
time. Material objects are spread out in time analogously to how they are
spread out in space.

One advantage of the temporal-parts account of material objects is that
(at least in some cases) it allows one to deny the crucial identity claims in
puzzles about material objects that raise the PMC without thereby admitting

spatially coincident objects. To see why, consider again the premises of SOT. The temporal-parts approach to the PMC allows one to reject premises 1, 3, and 5 of SOT on the grounds that these premises assume that aggregates and ships are material objects that *endure* through time (IMO3), that is the temporal-parts theorist does not think that ships and aggregates are material wholes that have different properties at different times. For the temporal-parts theorist, the Ship of Theseus has temporal as well as spatial parts. Therefore, the original ship at t and the continuous ship at $t+1$ are at most only temporal parts of the Ship of Theseus, and since one proper part of an object is not identical to another proper part (e.g. the left side of my body is not identical to my right), premise 5 of SOT is false according to the temporal-parts theorist. Note also that the temporal-parts theorist rejects crucial identity claims in SOT *without* thereby admitting the possibility of spatially coincident objects. Thus, the advocate of the temporal-parts view can solve the PMC by denying the identity claims in the puzzles that raise the PMC without denying IMO5. Of course, from the perspective of common sense, the price paid by the temporal-parts theorist is just as high as that paid by the advocate of the standard account. The temporal-parts theorist also solves puzzles about material objects by denying a common-sense belief about material objects, namely, that the material objects of our everyday experience are material wholes and not simply temporal phases of material objects.

Allan Gibbard has offered what many take to be a devastating counterexample to the temporal-parts account of material objects by way of his Lumpl/Goliath puzzle, a puzzle about material objects designed specifically to cause problems for the temporal-parts account.[18] In this puzzle we are to suppose that Goliath is a statue and Lumpl is the portion of clay that constitutes Goliath. We are also to suppose that Goliath and Lumpl have come into existence at the same time and are both destroyed at the same time. Because in this case Goliath and Lumpl would have the very same set of temporal parts, the advocate of the temporal-parts view can't deny the identity of Goliath and Lumpl in virtue of their having different temporal parts. But this leaves us with the contradiction that Goliath and Lumpl are both identical (because they are composed of the very same parts) and non-identical (since statues and pieces of clay have different essential properties). Thus, the Lumpl/Goliath puzzle raises the PMC in such a way that the temporal-parts view cannot dissolve it. At most, therefore, the temporal-parts view is only a *partial* solution to the PMC.[19]

The *relative-identity view* solves the PMC by denying that, necessarily, identity is transitive (IMO6). However, there are at least three versions of this view in the contemporary philosophical literature. One kind of relative-identity theory denies that identity claims translate across possible worlds. Gibbard has proposed that, though x is identical to y in world W, there is a possible world where x and y both exist and x is not identical to y. The Lumpl/ Goliath puzzle gives this view at least an apparent plausibility.[20] It stipulates that Goliath and Lumpl are composed of the same atoms and enjoy the very same histories in world W. They are thus identical in W. But Goliath and Lumpl might have had different histories, e.g., say Lumpl pre-exists Goliath in world $W1$, and therefore Lumpl and Goliath are not identical in that possible world. Therefore, although Goliath and Lumpl are identical in W, they are not identical in $W1$. But mustn't two objects be identical in all possible worlds in which they both exist if they are identical in one? Consider the following argument. If $x = y$ in world W, then x and y have all of the same essential properties in W. But one of the properties that x has in W is the modal property *necessarily being identical to x*, since everything is necessarily identical to itself. But if y is identical to x in W, then y in W also has the property *necessarily being identical to x*. Thus, it follows that if $x = y$ in W, then, necessarily $x = y$ (that is, $x = y$ in every possible world in which x and y exist). Gibbard gets around this argument by denying that objects have essential properties. If objects don't have essential properties, then they don't have properties such as *necessarily being identical to x*. In other words, IMO6 assumes that objects have essential properties. But they don't according to Gibbard. Therefore, IMO6 is to be rejected. Identity is relative to possible world.

A second kind of relative-identity theory proposes, more radically, that identity is relative to time.[21] Making identity relative to time entails that x might be identical to y and non-identical to y in the same possible world, albeit at different times. The Ship of Theseus puzzle in particular suggests that this might be the right way to think about the identity relation. Though there are perhaps two different ships that have a claim on being related to the original ship of Theseus at $t + 1$ (the reconstructed ship and the continuous ship), it is clearly the case, one might argue, that there was only one self-identical ship prior to the original ship's losing any of its parts. Thus, though the reconstructed ship and the continuous ship are clearly not identical with each other at $t + 1$ (since they exist in two different places, respectively), there is a time (namely, at t) when they are identical with one another. The metaphysician who thinks that identity is a *temporary* relation might tell such a story.

Finally, Peter Geach has famously argued that identity is relative to *sortal*: x and y may be the same F, but not the same G.[22] For example, consider the case of Chris (a human being) and the different aggregates of fundamental particles that constitute Chris at two different times – call the aggregate that constitutes Chris at t 'Aggregate1' and the aggregate that constitutes Chris at $t + 1$, 'Aggregate2'. With these objects in mind, a paradox can be generated. Aggregate1 is identical to Chris at t – otherwise we have to admit the possibility of two spatially coincident objects, namely, that Chris and Aggregate1 are two different objects existing in the same place at the same time. By parity of reason, Aggregate2 is identical to Chris at $t+1$. Chris at t is identical to Chris at $t+1$ (by stipulation). From these identity claims we can conclude that Aggregate1 is identical to Aggregate2, by way of the necessity of the transitivity of identity (necessarily, if $x=y$, and $y=z$, then $x=z$). But Aggregate1 and Aggregate2 are not identical to one another (by stipulation these aggregates have different parts and aggregates x and y are identical only if x and y have all and only the same component parts). A paradox!

The sortal relativist calls into question all of the identity claims made above, by calling into question the necessity of the transitivity of identity (IMO6). For example, what does 'Chris is identical to Aggregate1' really mean? The sortal relativist thinks that this claim is most perspicuously interpreted as follows: 'Chris and Aggregate1 are the same parcel of matter (but not the same human being).' So too for the claim 'Chris is identical to Aggregate2'. But in contrast, the sortal relativist takes the claim 'Chris at t is identical with Chris at $t+1$' to mean, 'Chris at t and Chris at $t+1$ are the same human being (but not the same parcel of matter)'. Given the ways the sortal relativist has disambiguated the identity claims above, one can't conclude that Aggregate1 is the same parcel of matter as Aggregate2 – that would require that 'Chris at t is identical to Chris at $t+1$' means 'Chris at t is the same parcel of matter at Chris at $t+1$', which it doesn't. The sortal relativist thinks that IMO6 – since it assumes that one can make absolute identity claims – is an assumption that should be rejected. So the sortal relativist rejects that one can make absolute identity claims such as '$x=y$'. Rather, all claims such as '$x=y$' have to be interpreted as 'x is the same F as y'.[23]

Another way of solving the PMC is to deny the existence of at least one kind of object in every puzzle that raises the PMC. This solution to the PMC might be referred as the *eliminativist view*. The eliminativist thus denies IMO2. For example, one way to solve SOT is to argue that there are no such things as ships, *or* that there are no aggregates (of planks). Many of the objects thought to exist by the ontological pluralist have a questionable ontological status according the eliminativist. For example, the ontological pluralist probably thinks that whatever ultimately constitutes an object – say, an aggregate of particles – is itself a compound material object. An eliminativist might reply that an aggregate of particles is not rightly described as one thing at all; 'aggregate' is a term that rather refers to a plurality of individual things.

Some eliminativists who accept IMO1 nonetheless come close to denying that there are any compound material objects at all. For example, the *mereological essentialist* is also a kind of eliminativist; specifically the mereological essentialist denies the existence of compound material objects that can survive changes with respect to their parts. The mereological essentialist thus denies IMO4. We intuitively think that a cat can survive the loss of her tail. The mereological essentialist denies this. But, more than this, every organism is constantly losing, gaining and replacing minute parts of itself. Thus, according to the mereological essentialist, no organism survives for very long, at least, not 'in the strict philosophical sense'. That is, we might treat tables, dogs and human bodies that lose parts as though they remain in existence as numerically the same object for pragmatic reasons (indeed, we must!). But, ontologically speaking, we would be wrong to consider such objects as remaining numerically the same through such changes.[24]

The advantage in taking up the eliminativist view as an answer to the puzzles that raise the PMC is the advantage that accrues to any eliminativist or reductionistic point of view: explanatory simplicity. Puzzles involving statues, lumps, aggregates and/or arbitrary undetached parts are not at all worrisome since such things are simply non-existent. Of course, a problem with this solution is that many of us have a strong intuition that garden-variety compound material objects are some of those objects that have the greatest claim on existence for us, since these are the objects that, as Baker puts it, 'actually matter to us'.[25]

The final way of solving the PMC I want to consider here is called the *dominant-kinds view*. First proposed by Burke,[26] and recently defended by Rea,[27] the dominant-kinds view might be considered a species of the eliminativist view. However, instead of rejecting the existence of whole classes of objects – e.g. saying that there are no such things as artefacts – the advocate of the dominant-kinds view thinks that, for every puzzle that raises the PMC, one of the *particular* objects mentioned in the puzzle does not have essential properties, and therefore does not exist.

Rea proposes that, for any puzzle that purportedly raises the PMC, one of the kinds thought to be present in that puzzle is merely a 'nominal' kind, and not a 'classificatory' kind. Because an object has the essential properties of its classificatory kind only, and not of just any nominal kind, the assumption (on the part of the standard account) that material constitution is a relation between two objects that have different essential properties is unfounded.

Consider a statue and the lump of clay that constitutes it. According to the standard account, this case involves two separate objects existing in the same place at the same time, since statues and lumps of clay have different persistence conditions, and therefore different essential properties. Melt a statue and it is certainly destroyed, but the same lump of clay might remain; break off a piece of the clay and a marred statue remains, but the lump of clay that constituted the statue is destroyed (although it is replaced by a new lump). In contrast, Rea and Burke suggest that there are not two objects here, but only one. Nevertheless, they acknowledge that the one object that exists belongs to two different kinds, but that one of these kinds is merely a nominal kind. When the lump of clay becomes a statue, the original lump is destroyed, and a new lump takes its place, one that is identical to a statue. Though the object satisfies both the kind *statue* and the kind *lump of clay*, the kind *statue* 'dominates' the kind *lump of clay*. As Rea puts it, 'I . . . deny that in saying that there is a lump of bronze in the region we are committed to the claim that there is something in the region that has the essential properties associated with the kind *lump of bronze*.'[28]

One implication of the dominant-kinds view that some find puzzling is the denial that things belonging to a kind always have the essential properties that we normally associate with that kind. If there really is a lump of bronze present where the statue is, how could it not have the essential properties of a lump of bronze? Furthermore, both Rea and Burke must work hard to show that there is not a 'which one' problem for their approach. Why should one say, for example, that the kind *statue* dominates the kind *lump of clay* in the example mentioned above? Why not rather say that *lump of clay* dominates the object in question instead of *statue*?

As Burke points out, the advantage of the dominant-kinds view consists in the fact that it 'dispense[s] with coinciding objects, but without relativising identity and without engaging in revisionist metaphysics, that is, without surrendering the elements of the standard account'.[29] In other words, Burke thinks that the dominant-kinds approach solves the PMC without rejecting the common-sense assumptions listed above (IMO1–6): there are such things as compound material objects (IMO1); there are many, many kinds of compound material objects (IMO2); material objects endure through time and change (IMO3), even changes with respect to their parts (IMO4); coincident objects are impossible (IMO5); and, necessarily, identity is transitive (IMO6).

Having offered a survey of the most influential contemporary approaches to solving the PMC, my examination of the problem of material constitution itself is now complete. I turn in Chapter 2 to offering detailed expositions of a few of the more sophisticated recent treatments of material constitution in the contemporary philosophical literature. These approaches to material constitution are also helpful for my purposes since each of them shares a number of key metaphysical commitments with Thomas Aquinas, whose views on material constitution I will be treating in detail in later chapters. The next chapter will therefore serve to prepare the reader for a detailed examination of a medieval philosopher's account of material constitution, one that – as I hope to show – has something to contribute to contemporary discussions of the PMC.

Notes

1. See, for example, Plutarch's *Life of Theseus*, book xxiii.
2. *Elements of Philosophy*, Concerning Body, ch. xi, 'Of Identity and Difference', section 7.

3. See, for example: Van Inwagen, P. (1981) 'The doctrine of arbitrary undetached parts', *Pacific Philosophical Quarterly* 62: 123–37; Heller, M. (1984) 'Temporal parts of four-dimensional objects', *Philosophical Studies* 46: 323–34; Burke, M. B. (1992) 'Copper statues and pieces of copper: a challenge to the standard account', *Analysis* 52: 12–17; Zimmerman, D. W. (1995) 'Theories of masses and problems of constitution', *Philosophical Review* 104: 53–110, and Rea, M. (1997a) 'Introduction' in Michael C. Rea (ed.), *Material Constitution: A Reader*, Lanham: Rowman & Littlefield, pp. xv–lvii.

4. Rea 1997a.

5. Baker, L. R. (2000) *Persons and Bodies: A Constitution View*, Cambridge: Cambridge University Press, p. 25.

6. I'll have occasion to mention other classical (and contemporary) puzzles about material objects in the sequel. In Chapter 6, I address a number of the classical and contemporary puzzles that raise the PMC in some detail.

7. My formulations of the classical and contemporary puzzles about material objects – formulations of the puzzles that explicitly raise the PMC – are heavily indebted to Heller 1984 and Van Inwagen 1981. For an alternative way to see the PMC, see Rea 1997a. Although I do not follow Rea's way of formulating the PMC, my understanding of these issues owes much to Rea's work.

8. Note that one might deny premise 1 of SOT and accept IMO5 by denying that objects endure through time (that is, by denying IMO3). I explain this solution and the problems that attend it in the sequel.

9. Following a custom of contemporary philosophers, I take a *proper* part of material object *x* to be a part of *x* that is not identical to *x*. In contrast, an *improper* part of *x* is that part of *x* that is identical to *x*.

10. I follow Burke (1992, p. 12) in speaking about the metaphysic of material objects that accepts the possibility of coincident objects as the 'standard account'. See also Burke (1997) 'Coinciding objects: reply to Lowe and Denkel', *Analysis* 57: 11–18. Here Burke explains the appropriateness of the title by noting 'its popularity and . . . its consistency with the metaphysic implicit in ordinary ways of thinking' (p. 11).

11. Burke 1992, pp. 12–13. For defences of the standard account, see: Wiggins, D. (1968) 'On being in the same place at the same time', *The Philosophical Review* 77: 90–5 (reprinted in Rea 1997b); Simons, P. (1987) *Parts: A Study in Ontology*, Oxford: Clarendon Press, and Lowe (1998) *The Possibility of Metaphysics: Substance, Identity, and Time*, Oxford: Clarendon Press. For a thorough list of recent advocates of the standard account, see Burke 1992, pp. 12–13 n. 1.

12. Burke 1992, p. 13.

13. Lewis (1986) *On the Plurality of Worlds*, Oxford: Blackwell, p. 252 (quoted in Baker 2000, p. 167).

14. Burke 1992, p. 13.

15. See, e.g., Wiggins 1968.

16. 1992, p. 14. For more development of this line of attack on the standard account, see Zimmerman 1995, pp. 87–91.

17. Baker 2000, p. 171.

18. Gibbard 1975.

19. Of course, there are other reasons aside from solving the PMC for accepting the view that material objects have temporal as well as spatial parts. For support of the view that material objects have temporal parts, see, for example: Heller 1984; Lewis 1986; and Sider, T. (2003) *Four-dimensionalism: An Ontology of Persistence and Time*, Oxford: Oxford University Press. For additional arguments against material objects having temporal parts, see, for example: Van Inwagen 1981 and Thomson, J. J. (1983) 'Parthood and identity across time', *Journal of Philosophy* 80: 201–20 (reprinted in Rea 1997b).

  Three-dimensionalism (captured by adherence to IMO3) seems to be the default position of most contemporary philosophers, and is also Thomas Aquinas's view (see, for example, In Phys. III, lec. 10, n. 373). Furthermore, I take the debate between four-dimensionalists and three-dimensionalists to be an exclusively contemporary one. Thus, it would be anachronistic to expect a defence of three-dimensionalism from Aquinas. Thus, given my interests in this work – articulating a common-sense metaphysic of material objects with the help of Aquinas – I do not treat four-dimensionalism as a serious option for solving the PMC in this work.

20. Gibbard 1975.

21. See, for example, Myro, G. (1997) 'Identity and time' in M. Rea (ed.), *Material Constitution: A Reader*, Lanham: Rowman & Littlefield, pp. 148–72.

22. Geach, P. (1967–8) 'Identity', *Review of Metaphysics* 21: 3–12.

23. Aside from denying the common-sense logical intuition that is IMO6, Rea also argues that Geach's view suffers the same problem as the temporal-parts view as far as solving the PMC is concerned: it cannot solve every puzzle that raises the PMC (1997a, pp. xlix ff.).

24. For arguments to this effect see, for example, Chisholm, R. (1998) 'Identity through time' in P. van Inwagen and D. Zimmerman (eds) *Metaphysics: The Big Questions*, Malden: Blackwell, pp. 173–85.

25. Baker 2000, p. 24.

26. Burke, M. (1994a) 'Preserving the principle of one object to a place: a novel account of the relations among objects, sorts, sortals, and persistence conditions', *Philosophy and Phenomenological Research* 54: 591–624.

27. Rea, M. (2000) 'Constitution and kind membership', *Philosophical Studies* 97: 169–93.

28. Rea 2000, p. 169.

29. Burke 1997, p. 11.

2

Three Contemporary Approaches to Solving the PMC

In Chapter 1 I described a problem (the problem of material constitution, or PMC) that arises for our common beliefs about compound material objects, namely, they appear to be logically inconsistent. Although I sketched some solutions to the PMC most often discussed by contemporary philosophers in Chapter 1, I now turn to examining in greater detail three contemporary views on the nature of compound material objects, those developed by Lynne Rudder Baker, Peter van Inwagen and Dean Zimmerman. These authors have contributed some of the more provocative arguments in the literature on material objects in recent years. But they also serve as useful interlocutors for explaining and evaluating Thomas Aquinas's views on material constitution. As we'll see, Aquinas's approach to material objects differs from all three of these authors in crucial ways. My hope is that by carefully investigating the views of some contemporary philosophers, we will be in a better position to make explicit the critical differences in philosophical assumptions made by Aquinas and our contemporaries concerning the nature of compound material objects. Getting clear on such differences will put the reader in a better position to evaluate the overall success of a Thomistic solution to the puzzles about material objects that raise the PMC.

Lynne Rudder Baker's constitution view

The first contemporary approach to the nature of compound material objects that I want to examine is one that Lynne Rudder Baker calls the *constitution view*.[1] One of the more important elements of Baker's constitution view is the following claim:

(CNI) Constitution is not identity.

An important entailment of CNI is that it is not necessarily the case that a material object x is identical to the aggregate of fundamental particles with which x is spatially coincident.

Baker offers a defence of CNI along the following lines.[2] Take a statue, such as Michelangelo's *David*. *David* is something constituted by a particular *piece* of marble. (Baker calls the piece of marble that constitutes *David*, 'Piece'.) What is the relation between *David* and Piece? One might think that the relation between them has to be identity, since both *David* and Piece are composed of just the same atoms. But Baker argues that the relation between *David* and Piece cannot be that of identity, since *David* and Piece have different essential properties. In contrast to Piece, *David* has the property *being related to an art-world* essentially. *David* exists only in virtue of an artist such as Michelangelo sculpting it and a community of rational beings appreciating it as a work of art. Whereas Piece exists in possible worlds that lack communities of artists, art critics, etc., *David* does not. For Baker, the case of *David* and Piece is a paradigm instance of material constitution.[3] Since two things that have different essential properties cannot be identical, Baker concludes that constitution is not identity.

The nature of the constitution relation

But if constitution is not identity, what is it? Baker takes constitution to be a kind of *unity* relation.[4] As she remarks, 'constitution is an intimate relation – almost as intimate as identity, but not quite' (2000, p. 179). Thus, although the *relata* of the constitution relation are not to be identified with one another, neither should they be counted as 'two independent individuals' (2000, p. 31). Here we can see Baker attempting to distinguish the constitution view from what I take to be its philosophical cousin – the standard account of material constitution, a view with which the constitution view shares the conviction that constitution is not identity.

Baker offers several reasons for thinking that Piece and *David* are not two independent material objects as advocates of the standard account suppose (2000, p. 31). First of all, they are spatially coincident objects. *David* does not exist separately from Piece. At minimum, *David* is constituted by Piece at every time in which *David* exists. In addition, as Baker notes, *David* and Piece have many of the *same* properties, e.g. those having to do with 'size, weight, color, smell, and so on' (2000, p. 31). Second, some of *David*'s aesthetic properties depend on Piece's physical properties, e.g. '*David*'s pent-up energy depends on, among other things, the way that the marble is shaped to distribute the weight' (2000, p. 31). Finally, *David* does not have Piece as a proper part. As Baker explains: 'For pretty clearly, *David* is not identical to Piece plus some other thing' (2000, p. 31). For these reasons Baker claims that *David* and Piece are not two independently existing individual objects as the standard account supposes.

So in what sense are material objects spatially coincident whenever they come to exist in the relation of material constitution according to Baker? The standard account of coincident objects suggests that two separately existing things can – and do, in cases of material constitution – spatially coincide in the following strong sense: whenever Piece constitutes *David*, Piece and *David* are two separate, individual material objects. Baker understands the constitution view to be committed to spatial coincidence in a weaker sense than the standard account. For example, she says

> Since a large part of my task is to distinguish constitution from identity, I ... [emphasize] ways in which *x* and *y* are distinct if *x* constitutes *y*. But too much emphasis on their distinctness would be misleading: for, as we see in the examples of, say, a statue and a lump of clay that forms it, *x* and *y* are not separate, independently existing individuals. Again: I want to make sense of constitution as a third category, intermediate between identity and separate existence. (2000, p. 29)

Elsewhere, she says

> If *x* constitutes *y* at a certain place and time, then there is a unified individual at that place at that time, and the identity of that individual is determined by *y* ... The identity of the constituting thing is submerged in the identity of what it constitutes. As long as *x* constitutes *y*, *y* encompasses or subsumes *x*. (2000, p. 33)[5]

In Baker's view, *David* and Piece are not two separately existing, spatially coincident objects. However, in that case it is not yet clear just how *David* and Piece are supposed to be spatially coincident according to Baker. What we need is a clear account of the ontological status of *David* and Piece when Piece constitutes *David*. One might think that when *David* exists, Piece goes out of existence altogether – Piece's existence being replaced by that of *David*. Of course, this can't be Baker's view – though it is the view of some contemporary metaphysicians[6] – since in that case *David* and Piece wouldn't be spatially *coincident* at all. A better suggestion would be that, when *David* comes to exist, Baker thinks that Piece continues to exist, but does so in a sense different from the way that *David* exists. Although Baker never explicitly argues in this way, this interpretation of the relation between *David* and Piece is supported by Baker's use of metaphors such as 'subsumed', 'encompassed', and 'submerged' to describe the constituting thing's relation to the constituted thing. But in the end, it just is not clear what Baker takes the ontological status of Piece to be when Piece constitutes *David*.

Nevertheless, Baker attempts to distinguish her constitution view from the standard account by appealing to the notion that individual things can have properties derivatively and non-derivatively. For example, in Baker's view Piece has many of its properties only in virtue of constituting something such as *David*, e.g. the property *being a statue*.[7] *David* also has many of its properties only in virtue of being constituted by Piece, e.g. the property *being made of marble* (2000, p. 47). Thus, *David* and Piece each have the property *being a statue*, but they have this property in different ways. *David* has the property *being a statue* non-derivatively, or *independently* of its constitution relation to Piece (or anything else). But when Piece constitutes *David*, Piece also has the property *being a statue*, although it has that property only derivatively, or only in virtue of constituting *David* (or something like it).

Now according to the standard account, there is no sense in which Piece has the property *being a statue* or *David* has the property *being a piece of marble*. The reason that the advocate of the standard account maintains this view has to do with a worry about coincident objects *belonging to the same kind*. In other words, if Piece too is a statue (where the 'is' here is the 'is' of predication), when Piece constitutes *David*, that would mean there were two *statues* existing in the same place at the same time, and this is a kind of spatial coincidence that even the advocate of the standard account takes to be absurd.[8]

Baker argues that the constitution view does not have the implication there are two statues existing in the same place at the same time when Piece constitutes *David*. Derivative properties are not 'additive' in this way. Baker explains:

> The reason that derivative properties are not 'additive' is that *there is nothing to add*: *x*'s having F derivatively is nothing other than *x*'s being constitutionally related to something that has F non-derivatively. .... If *x* and *y* have constitution relations and *x* is an F, then *x* is the same F as *y*. (2000, p. 177)

This view of derivative properties is a consequence of her view that constitution is a type of unity relation. As Baker explains in another place, if *x* constitutes *y* at *t*, *there are not two things* existing where *x* and *y* are, as the standard account would seem to suggest, rather, there is one thing, *y*, that is constituted by *x*.

> For when *x* constitutes *y*, there is a unitary thing – *y*, as constituted by *x* – which is a single thing in a sense that I shall try to make clear. As long as *x* constitutes *y*, *x* has no independent existence. If *x* continues to exist after the demise of *y*, then *x* comes into its own, existing independently. But during the period that *x* constitutes *y*, the identity of 'the thing' – *y*, as constituted by *x* – is determined by the identity of *y*. (2000, p. 46)

Thus, according to the constitution view, Piece and *David* derive properties from each other, and this marks that view off from the standard account. Like the standard account, the constitution view holds that Piece and *David* are not identical, and that their non-identity can be most clearly seen in virtue of the fact that they have different essential properties. However, in contrast to the standard account, Baker takes the constitution relation to be a relation 'of genuine unity' (2000, p. 28). Baker thus takes the relation of constitution to be an ontological category 'intermediate between identity and separate existence' (2000, p. 29).

The implied ontology of the constitution view

Baker takes both *relata* of the constitution relation to be concrete individual things rather than properties, portions or stuffs (2000, pp. 33–4).[9] But what sorts of things count as individual material objects according to Baker? Baker has the view that every individual material object belongs to a single 'primary kind'. For Baker, a primary kind is a substance-sortal that answers the question for any given concrete individual object, 'what is that thing, most fundamentally?' (2000, pp. 39–40). A thing may belong to a number of different kinds, but a thing's primary kind determines its persistence conditions (2000, p. 170 n. 5). This is because the primary kind of a thing expresses, as I have said, what something is most fundamentally. It answers the question, in the most precise way possible, 'what is it?' For example, there is a white object in front of me. Although it is true that the object does have the property *being white*, its whiteness does not tell me what that thing is most fundamentally – indeed, the proof of that is that the object can survive the loss of the property *being white*. Say the object in question is also an animal; even its animality does not explain as precisely as possible exactly *what* it is (there are lots of different kinds of animals; or, the object may be an animal only contingently). Rather, this object is most fundamentally a *person*, and this description gets to the 'core', so to speak, of what this object is.

As Baker recognizes, being able to tell the difference between mere property acquisition and primary-kind property acquisition in all cases would require that one have a theory of primary kinds, that is, a theory which would 'provide a principled way to distinguish between cases ... in which an object merely acquires a property and cases ... in which a new entity comes into existence' (2000, p. 40). Admitting that she does not have such a theory, Baker wants nevertheless to say that certain 'circumstances' function as sufficient conditions for an object x constituting an object y, where x and y are non-identical and belong to different primary kinds. As she states: 'when various things are in various circumstances, new things – new kinds of things with new kinds of causal powers – come into existence. When a piece of cloth is: in certain circumstances, a new thing, a flag, comes into existence' (2000, p. 20).

As is evident from Baker's example of a flag and the piece of cloth that it constitutes (think also of *David* and Piece), the sorts of 'circumstances' that bring about a case of material constitution may include relations to things extrinsic to the constituting and constituted thing. For example, Baker thinks that *David* has *being a statue* as its primary-kind property. But the property *being a statue* is one that statues have only in virtue of their having the appropriate sorts of relations to communities of persons that recognize and value physical objects as works of art, and these are relations that are certainly extrinsic to the atoms that compose the aesthetic object.

But what distinguishes circumstances that issue in a case of material constitution from those that merely issue in a case of property acquisition? Why do certain of Piece's extrinsic relations cause it to come to constitute an object such as *David* while others remain ordinary extrinsic relations such as *being to the left of another piece of marble*? Baker suggests we adopt the following principle: 'If *x* constitutes *y*, then *y* has whole classes of properties that *x* would not have had if *x* had not constituted anything' (2000, p. 41). In offering applications of this principle, Baker enlists the following helpful examples:

> [An] anvil acquires the property of being a doorstop by our enlisting a physical property of the anvil – its heaviness – for a special purpose: to hold open the barn door. The use of the anvil as a doorstop does not bring about instantiation of whole classes of properties that anvils *per se* do not have. On the other hand, *David* has many causal properties of different kinds that Piece would not have had if Piece had not constituted anything. (2000, p. 41)

Given such a principle for distinguishing genuine from non-genuine cases of material constitution, one might still wonder how pervasive Baker herself takes the constitution relation to be. We have seen that most of her examples involve objects whose existence depends upon certain extrinsic relations obtaining. But Baker is also quite clear that there are constituted things belonging to primary kinds in virtue of their intrinsic properties and not as a result of some extrinsic relation. Indeed, Baker takes all human persons and all organisms to be prime examples of constituted things 'whose primary kinds are in fact determined by [their] intrinsic properties' (2000, p. 170). Given that Baker also takes Piece itself to be something constituted (of an aggregate of fundamental particles), this means that she considers that pieces of stuff, organisms, human persons and artefacts are all examples of constituted things. Thus, as Baker suggests, according to the constitution view, the constitution relation is a 'ubiquitous relation' (2000, p. 9).

As we saw above, Baker takes the *relata* of the constitution relation to be two individual concrete things. But what sorts of objects count as concrete things according to Baker? Examples of classes of *concreta* for Baker include aggregates, pieces of some stuff-kind, and objects such as atoms, molecules, organisms, artefacts and artworks (2000, pp. 45–6).

Let me say something about how Baker understands the items on this list that are a little out of the ordinary, namely, aggregates and pieces of some stuff-kind. There are at least two different ways of understanding the meaning of 'aggregate'. According to one account, 'aggregate' refers to any number of individual things greater than one (call them 'the *x*s'), where there are no particular restrictions on the spatial proximity or the causal relations that obtain between the *x*s. Thus, on this account of 'aggregate', an aggregate of things includes *any* two or more individual things. My car and my favourite book count as an aggregate on this account. Call this the 'universalist' understanding of 'aggregate'. According to a second account, 'aggregate' refers to the *x*s just in case they are spatially proximate to one another or, more precisely, enter into causal relations with one another such that the *x*s are touching or bonded to one another. Call this the 'bonding' understanding of 'aggregate'. A pile of rocks or a statue-shaped group of iron atoms would be good examples of aggregates according to the bonding account. What both accounts of 'aggregate' have in common is the following uncontroversial condition for aggregate identity: aggregates *a* and *b* are numerically the same only if *a* and *b* are composed of the same set of parts.

Baker clearly endorses the universalist account of aggregates. For example, at one point she speaks of an existent aggregate of water molecules that is 'scattered throughout the universe' (2000, p. 172).[10]

Given that she employs 'aggregate' in accordance with the universalist account, what sort of ontological status does Baker grant to aggregates? Some things Baker says give her reader the impression that aggregates are not concrete individual things. In explaining the way she uses the term 'aggregate' Baker states in one place, 'I use the term "aggregate" to singularize a plurality of things' (2000, p. 215). This would seem to suggest that 'aggregate' is a term of convenience – it makes talk about the objects that ultimately compose a compound material object easier. Since Baker understands 'aggregate' according to the univeralist account, the view that aggregates are logical fictions would save her from thinking that any random group of material objects has a mind-independent ontological status, a view that is counter-intuitive, to say the least.

But Baker takes the constitution relation to have individual things as its *relata*. This is clearly a metaphysical claim on Baker's part, and so if *x* constitutes *y*, then both *x* and *y* must be individual things. But as Baker often reminds us, 'an aggregate of *x*s (as opposed to the *x*s) can . . . be a *relatum* of the constitution relation' (2000, p. 215).[11] These assertions would seem to suggest that Baker thinks of aggregates as individual things after all. I do not know how to synthesize Baker's apparently disparate views on the ontological status of aggregates. Baker seems to need aggregates to be individual things, but in other places she (I think, rightly) denies the counter-intuitive view that aggregates – at least understood according to the universalist account – are real individual things.

In addition to asking questions about the ontological status of aggregates of objects, philosophers interested in material objects also speak about portions and pieces of stuff-kinds. Baker thinks that *portions* of a stuff-kind are not, while *pieces* of some stuff-kind are, ontologically significant (2000, pp. 33–4 n. 19). But just what does Baker take a *piece* of some stuff-kind to be? A good way to get clear on Baker's understanding of a piece of stuff-kind is to compare what she says about pieces with what she says about portions.

In distinguishing portions and pieces, Baker notes that they have different persistence conditions where their relations to their parts are concerned. As she makes clear in an extended footnote, Baker takes a portion of, say, cotton to be such that it would go out of existence if it gained or lost a part (2000, p. 33 n. 19). This is not an idiosyncratic view of the term 'portion'. As we saw in Chapter 1, many philosophers think of *portions* as being essentially related to their proper parts.[12] Unlike portions, Baker takes pieces of some stuff-kind to be such that they can survive the loss of a proper part. That Baker thinks this is so is clear from her statement that Piece exists prior to being quarried (2000, p. 29 n. 9). Thus, Baker thinks that Piece survives the loss of those parts that are invariably separated from it between the time it existed in a marble quarry and the time it comes to constitute *David*. Pieces of some stuff-kind are not essentially related to their parts according to Baker.

But what are the identity conditions for pieces of some stuff-kind according to Baker? Are pieces of some stuff-kind like living organisms in being essentially related to some of their parts? If so, which of Piece's parts are essential to Piece's continued existence? Of course, there are fairly clear answers to this sort of question in the case of living things. A living thing is essentially related to those of its parts without which it cannot perform, or have the capacity to perform, its distinctive function, and/or stay alive. Thus, if the distinctive function of some plant is to reproduce, one might argue that that plant goes out of existence whenever it loses the capacity to reproduce itself. However, these lines of reasoning are harder to apply to something such as Piece. Presumably, Piece *cannot* survive just any kind of loss of its parts. If Piece were blown into tiny bits, its parts thereby being spread throughout an area with a radius of one hundred feet, surely Baker would not want to say that Piece continues to exist. Piece goes out of existence if it is blown to smithereens. Which of Piece's parts are integral to its continuing to perform its function? Baker nowhere answers the questions I have been raising here, and they would seem to be difficult to answer. No wonder that most metaphysicians have opted instead to speak of portions of some stuff-kind, whose identity conditions are clear: a portion of some stuff-kind stays in existence just as long as all of its parts remain contiguous with (or are somehow bonded to) one another (or, on a 'universalist' understanding of portions, a portion of some stuff-kind stays in existence as longs as its parts stay in existence).

Nevertheless, it is clear that many of us believe in the existence of objects such as Piece. If I throw a rock, and it skips across the ground a number of times before coming to a rest, chances are that it will have lost a proper part or two (or two hundred) in the process. Nevertheless, when I go to pick up that rock, I consider it to be the same numerical rock as before. Baker's view that there are *pieces* of stuff in the world definitely has some intuitive appeal.

Baker's constitution view and the PMC

Although Baker never explicitly addresses the PMC, it is clear that she would *not* respond to the puzzles that raise the PMC by advocating the existence of temporal parts (2000, p. 29 n. 8), by advocating mereological essentialism, or by denying the necessity of the transitivity of identity (2000, p. 31). As an ontological pluralist, Baker also accepts IMO1 and IMO2.

Of the assumptions that lead to the PMC, Baker clearly rejects IMO5. According to the constitution view, if x constitutes y, it might be the case that x and y have all of the same parts, and yet x and y are not identical (2000, p. 181).[13] Thus, to take one example, Baker would solve the Ship of Theseus puzzle by denying that a ship is identical to the aggregate of planks that constitutes it, while also admitting that the ship and the aggregate are spatially coincident material objects.

Nonetheless, Baker has tried to make her acceptance of spatially coincident objects (i.e. the rejection of IMO5) more palatable by arguing that the constitution relation does not have two separately existing material objects as its *relata* – as in the case of the standard account of material constitution. Instead, she thinks that the constitution view has it that constitution is a type of *unity* relation weaker than identity but stronger than the relation between separately existing individual objects. The availability of such a *sui generis* relation between identity and separate existence is crucial to Baker's solution to puzzles that raise the PMC, since in her view it quiets concerns raised by those who balk at the possibility of spatially coincident material objects.

Peter van Inwagen on composition, eliminativism and the PMC

Unlike Baker, Van Inwagen has explicitly acknowledged the PMC, and has offered solutions to the puzzles that raise it in a number of his works.[14] Since Van Inwagen accepts the PMC, he recognizes that intuitions IMO1–6 constitute an inconsistent set, and that he must therefore reject one of the common-sense assumptions typically made about compound material objects. Furthermore, he has gone on record that he accepts all of the intuitions about compound material objects that generate the PMC, save IMO2.[15] Since he accepts intuitions IMO1, IMO3, IMO4, IMO5 and IMO6, and that IMO1–6 are an inconsistent set, Van Inwagen rejects IMO2 for all the puzzles that raise the PMC. Thus, Van Inwagen is an eliminativist when it comes to the PMC: for every puzzle that raises the PMC, there is at least one object talked about in that puzzle that is merely a virtual object.

Van Inwagen's most thorough defence of his rejection of ontological pluralism comes in 1990a. I want now to turn to summarizing the central argument of that work, an argument that comes in answering a question he calls 'the Special Composition Question'.

Van Inwagen, the Special Composition Question and the PMC

In a book-length treatment of the metaphysics of material objects, Van Inwagen argues that there are a lot fewer kinds of material beings than philosophers often suppose.[16] His thesis: the only material beings that exist are living organisms and the fundamental particles that compose them (1990a, p. 98). As Van Inwagen notes, an important consequence of such a theory of material objects is that there are no artefacts, non-living compounds or arbitrary undetached parts (e.g. hands, heads and hearts). Van Inwagen calls this consequence of his theory of material objects, 'the Denial' (1990a, p. 1).

Why would one accept a theory of material objects that has as a consequence the Denial? As I have already suggested, the PMC itself provides a reason for the development of a leaner ontology of this sort. In Van Inwagen's view, of all the intuitions that together generate the PMC – IMO1, IMO2, IMO3, IMO4, IMO5 and IMO6 – the rejection of the assumption that there are material objects such as aggregates, non-living things, arbitrary undetached parts and artefacts (IMO2) is easiest to defend (1990a, pp. 3–18).

Van Inwagen defends the Denial in the context of offering an answer to what he calls 'the Special Composition Question' (SCQ).[17] Although Van Inwagen formulates the SCQ in a variety of different ways, we can take what he calls a 'practical' formulation of the SCQ as a representative example: 'Suppose one had certain (non-overlapping) objects, the *x*s, at one's disposal; what would one have to do – what *could* one do – to get the *x*s to compose something?' (1990a, p. 31).

Before taking a look at Van Inwagen's argument for his answer to the SCQ, it will be useful for my purposes to say something about one additional assumption behind his arguments in 1990a: his acceptance of philosophical *atomism*. The difference between atomism and non-atomism has to do with the question whether or not the ultimate constituents of material things are simples (entities that can't be divided). Say for the sake of argument that water is an ultimate constituent of all things, that is that portions of water cannot be divided so as to result in a kind of thing that is not water. The non-atomist would think that *every* part of the water in a glass of water is a portion of water that can itself be divided into portions of water, *ad infinitum*. Van Inwagen thinks that current theories of physics favour the view that the ultimate constituents of compound material objects are simples (1990a, p. 72). So, as Van Inwagen relates, 'I assume that every material thing is composed of things that have no proper parts: "elementary particles" or "mereological atoms" or "metaphysical simples"' (1990a, p. 5). Van Inwagen thus takes matter to be 'ultimately particulate', (1990a, pp. 5, 15)[18] or ultimately atomistic in character.[19]

The SCQ: some proposed solutions

So when do the xs compose something? Common sense might suggest that contact is a sufficient condition for composition. Van Inwagen terms this answer to the SCQ, 'Contact'. According to Contact, 'to get the xs to compose something, one need only bring them into contact; if the xs are in contact, they compose something; and if they are not in contact, they do not compose anything' (1990a, p. 33). Van Inwagen offers some convincing arguments to the effect that Contact cannot be the right answer to the SCQ. First, even if Contact would seem to be true where the *visible* parts of garden-variety material objects were concerned, it would not come out true for the *invisible* parts of *those same* garden-variety material objects. As twentieth-century physics has taught us, 'the elementary particles that compose a given material object are not in contact' (1990a, p. 34). Thus, if physical contact between objects is a necessary condition for composition, then there really are no compound material objects.

A second argument lists a counter-example to Contact's claim that physical contact is a sufficient condition for material composition. Imagine that two human beings shake hands at time *t*, thus coming into contact at *t*. Our intuitions strongly suggest that these two human beings do not come to compose something at *t* that has two human beings as its proper parts (1990a, p. 35). Thus, being in contact cannot be a sufficient condition for the xs coming to compose something.

Next consider three possible 'bonding' solutions to the SCQ. Each of these answers suggests that a sort of physical bonding relation – each sort of physical bonding relation being progressively stronger than the last – is sufficient to bring about a case of genuine composition. For example, the xs are *fastened* if and only if the xs are in contact and . . . are so arranged that, among all the many sequences in which forces of arbitrary directions and magnitudes might be applied to either or both of them, *at most only a few* would be capable of separating them without breaking or permanently deforming or otherwise damaging either of them. (1990a, p. 56)[20]

According to the Fastening answer to the SCQ, the *xs* compose an object if and only if the *xs* are fastened (or there is only one of the *xs*). A second bonding-type answer to the SCQ, Cohesion, has it that the *xs* compose an object if and only if they *cohere* to one another (or there is only one of the *xs*). Objects 'cohere' if and only if 'they can't be pulled apart, or even moved in relation to one another, without breaking some of them' (1990a, p. 58).[21] Finally, objects are 'fused' if and only if 'they melt into each other in a way that leaves no discoverable boundary' (1990a, p. 59).[22] Thus, Fusion is an answer to the SCQ that says the *xs* compose an object if and only if they are fused. Van Inwagen argues that none of these bonding relations provides a sufficient condition for an occurrence of composition, since our intuitions suggest that human beings that are bonded in any of these three ways would not come to compose an object that had human beings as proper parts.

There are other possible *bonding* solutions to the SCQ. To take Van Inwagen's most interesting example, a 'Series' style answer to the SCQ posits that different kinds of objects have different necessary and sufficient conditions for their coming to compose an object. For example, one might suggest that *non-living* things compose an object just in virtue of being bonded to one another in some way A, but the fact that two living things are A-bonded does not entail that those living objects in fact compose an object (1990a, p. 63). Van Inwagen formulates Series as follows: '$(\exists y$ the *xs* compose $y)$ if and only if the *xs* are F_1 and stand in R_1, or the *xs* are F_2 and stand in relation R_2, or . . . , or the *xs* are F_n and stand in relation R_n' (1990a, p. 63).

Although Series avoids the criticisms levelled at Fastening, Cohesion and Fusion, Van Inwagen thinks that it nevertheless fails to be a successful answer to the SCQ. This is because he expects a proper answer to the SCQ to take the position of the Nihilist seriously. (In this context 'Nihilism' picks out the ontological view that there are no compound material objects.) A Series-style answer to the SCQ simply consists of a list of the different compound material objects that are thought (by the Series-style advocate) to exist, as well as a list of the different relations that are thought (by the same Series-style advocate) to ensure the existence of such objects. In order to avoid begging questions against the Nihilist, Van Inwagen proposes that the right-hand side of a proper response to the SCQ cannot itself contain any terms whose meaning assumes the existence of composed objects of some sort.[23] For example, contrast Van Inwagen's formulation of Contact with a Series-style answer. Contact suggests that the *x*s compose something if the *x*s are in contact. The right-hand part of this response to the SCQ does not contain any reference to a composed object; rather, it merely offers a sufficient condition for the *x*s actually composing an object, where 'the *x*s' need not be construed as composed object themselves. But the Series-style response simply reports that certain kinds of relations supposedly bring about certain kinds of composed objects. Unlike the Series-style answer, Contact is subject to counter-example, whether by a Nihilist or someone else, and so it is not a question-begging response to the SCQ. Of course, as we have seen Van Inwagen thinks Contact fails as an answer to the SCQ for other reasons.

Van Inwagen's proposed answer to the SCQ

Van Inwagen argues that the correct answer to the SCQ must involve causal relationships between the *x*s, and not simply spatial relationships (1990a, p. 81). Specifically, his thesis on material composition is that the *x*s compose something if and only if 'the activity of the *x*s constitutes a life (or there is only one of the *x*s)' (1990a, p. 82).

Let me spend some time explaining what the components of this claim mean before showing how Van Inwagen attempts to justify it. By 'activity', Van Inwagen means to point to 'no more than a way of talking about the changes that the *x*s undergo' (1990a, p. 82). Although Van Inwagen does say that he is leaving the way that the activities of an object constitute an event 'at a more [or] less intuitive level' (1990a, p. 82), it seems that he owes us an explanation of his use of the term 'constitutes' in this context. This is because it is natural to think that 'constitutes' is either a synonym for 'composes' or, at the very least, is similar to 'composes' in meaning. In fact, given Van Inwagen's sensitivity to the employment of mereological terms in the right-hand side of an answer to the SCQ – that is terms that might imply that some mode of composition is in play – it is important that 'constitutes' not mean the same thing as 'composes'. However, in order to understand the implications of Van Inwagen's use of 'constitutes' here, it will be helpful first to get clear on Van Inwagen's use of the term 'life' in this context. What does it mean to say, 'the activity of the *x*s constitutes a *life*?'

Van Inwagen states that by 'life' he means to refer to 'the individual life of a concrete biological organism' (1990a, p. 83). In offering a descriptive explanation of what he takes the life of a living organism to be, he invokes a number of analogies. For example, Van Inwagen compares the activity of a life over time with that of a human institution, such as a club. Although a club's membership constantly changes over time, the activities or effects of that club nevertheless often continue in an uninterrupted fashion. The activity of a club is also made up of the combined and coordinated activities of the individual members of that club. Furthermore, the stability of an institution is not something static, but dynamic. As Van Inwagen writes,

> The stability of a typical social organization . . . is a dynamic stability. Some groups of people – one would not call them organizations – display a purely static stability. One might cite passengers trapped in a bus buried by a landslide (a case of stability imposed upon the members of the group by boundary conditions) [.] . . . The stability of our club in no way resembles the stability of the trapped passengers. (1990a, p. 85)

Finally, Van Inwagen is also quick to note that the constitution of a club is not 'an identifiable object, but rather a complex set of dispositions and intentions' (1990a, p. 84). Here Van Inwagen is saying, among other things, that a life is somehow related to, but not identical with, a living organism. Lives are events, or processes, whereas Van Inwagen takes living organisms to be individual material objects. But a living organism is an individual material object that undergoes a series of changes and developments for as long as it exists. Van Inwagen labels this entire series of changes the 'life' of a living thing.

Besides employing analogies to help explain what he takes the term 'life' to mean, Van Inwagen also suggests that lives have certain essential properties, one of which distinguishes them from all other types of events or processes. Lives are events that are 1. self-maintaining, 2. reasonably well-individuated and 3. jealous (1990a, pp. 86–9). Van Inwagen understands something to be self-maintaining if it retains its identity through time and change. Although lives are indeed so characterized, there are events that are self-maintaining without being lives, e.g. flames and waves. Thus, being self-maintaining is only a necessary but not a sufficient condition for being a life. In addition to being self-maintaining, lives are also well-individuated. If something is well-individuated according to Van Inwagen, that thing is able to be clearly distinguished from other things of its kind over the course of time and change. In contrast to lives, flames are not well-individuated. However, waves are also well-individuated; it is a simple enough procedure to follow the path of a wave through different constituting masses of water. But waves are not lives. However, lives also are jealous events, whereas waves are not. In contrast to two waves, which can be constituted by one and the same mass of water at the same time, 'it cannot be that the activity of the *x*s constitute at one and the same time two lives' (1990a, p. 89).[24]

Let me now return to the issue of what Van Inwagen means when he says 'the activity of the *x*s *constitutes* a life'. Given what we have seen him say about the distinction between events and individual things – the latter but not the former being 'continuants' or 'substances' – Van Inwagen seems to be using 'constitutes' in a way that differs from 'composes'. He uses 'constitutes' to speak about a way that the processes undergone by the *x*s can be related to other processes undergone by a physical thing composed of the *x*s, whereas he uses 'composes' to denote a unity relation that has physical things themselves as its *relata*. Thus, the life-event of an organism is constituted by the activities of the cells that compose that organism. Because Van Inwagen's interest in 1990a is the nature of proper parthood with respect to physical objects (p. 20), his use of a term such as 'constitutes' in the right-hand side of his proposed answer to the SCQ does not entail that his answer is a circular one. (Although it does seem to raise a question that is at least analogous to the SCQ: 'when is it the case that the activities of the *x*s constitute something, e.g. a life?')

Van Inwagen defends his answer to the SCQ in a number of ways. One argument points out that Van Inwagen's proposed answer to the SCQ provides satisfactory solutions to the classic and contemporary puzzles about material objects that raise the PMC. Because there are no artefacts according to the proposed answer, there are no philosophical problems for artefacts. Thus, Van Inwagen solves SOT by denying that there are such things as ships (and aggregates). Van Inwagen solves other puzzles about material objects that raise the PMC in similar fashion.

However, as we have seen, there are many other principled ways of solving puzzles about compound material objects that raise the PMC that do not have as a consequence the Denial. So what other reasons does Van Inwagen have for defending an answer to the SCQ that has the Denial as a consequence? Van Inwagen offers a number of different arguments in defence of the Denial and his position on the SCQ. I want to focus on just one of these here: there are good reasons for thinking there are no such material objects as artefacts.[25]

Van Inwagen finds the following sort of argument against the existence of artefacts compelling. If someone takes an artefact such as a statue to exist (and thus to be something essentially different from the lump of clay that constitutes it), then 'the property *being a statue* is possessed by the lump whenever the lump is of the right shape' (1990a, p. 126). But, if someone takes this view, then a lump of clay is always spatially coincident at any time with some arbitrary object. For example, Van Inwagen suggests that someone might name a lump of clay shaped in an arbitrary fashion a 'gollyswoggle'. It stands to reason that if an artefact comes into existence simply because a lump takes on a certain shape, then a gollyswoggle comes into existence whenever a lump becomes gollyswoggle-shaped. Van Inwagen continues

> if you can make a statue on purpose by kneading clay, then you can make a
> gollyswoggle by accident by kneading clay. But if you can make a gollys-
> woggle by accident by kneading clay, then you must, as you idly work the

clay in your fingers, be causing the generation and corruption of the members of a compact series of objects of infinitesimal duration. That is what seems incredible. (1990a, p. 126)

Given the implication that merely changing the shape of a mass of clay brings about 'a compact series of objects of infinitesimal duration', an implication that he thinks follows from asserting that there are such things as artefacts, Van Inwagen rejects that there are such things as artefacts.

Van Inwagen's answer to the SCQ as a defence of his answer to the PMC

Van Inwagen's basic approach to the PMC is eliminativist; puzzles about material objects can be solved by way of denying the existence of certain objects assumed to exist in each puzzle that raises the PMC. Although Van Inwagen does defend his answer to the SCQ by pointing out that it offers a principled way of handling the puzzles that raise the PMC, Van Inwagen also offers other defences of his answer to the SCQ that have nothing at all to do with the puzzles about material objects that raise the PMC. If these latter arguments are sound – arguments such as the argument against artefact existence outlined above – they offer support for Van Inwagen's eliminativist approach to the PMC. As we have seen, there are a variety of ways of solving the PMC – at least six – one for each of the intuitions (IMO1–6) that together appear to constitute an inconsistent set. But assuming that one has to deny one of these six intuitions to solve the PMC, why should we reject one rather than another? As we have seen, Lynne Rudder Baker thinks it is clear that there are such objects as pieces of marble (a non-living compound material object) and artefacts, and this leads her to reject a different intuition about material objects from Van Inwagen. Who's right, Baker or Van Inwagen? A complete response to the PMC provides a defence of why one intuition about material objects rather than another should be rejected. But this is just what Van Inwagen's proposed answer to the SCQ gives him. Since Van Inwagen can defend his proposed answer to the SCQ independent of any discussion of the PMC, his proposed answer to the SCQ – if it is indeed a good one – also provides him with a good reason for taking an eliminativist stance with respect to the PMC, i.e. rejecting IMO2 rather than another intuition about material objects.

Having taken a look at two fairly comprehensive approaches to the nature of compound material objects in the first two parts of this chapter, in the final part of this chapter I want to take a look at an argument developed recently by Dean Zimmerman. My discussion of Zimmerman's argument raises a number of conceptual issues that will be relevant to what is coming in later chapters – including the notion that the ultimate constituents of compound material objects are not particles but stuffs – and it also proposes a challenge to the viability of both of the accounts of material constitution that I have looked at so far.

The Zimmerman Argument

Dean Zimmerman has written a number of recent papers on problems having to do with the nature of material constitution.[26] In these papers he makes a powerful argument that the number of defensible positions concerning the constitution of material objects is smaller than many philosophers often suppose. In treating Zimmerman's work on issues relevant to material constitution, I confine myself to reconstructing this argument, one that I will refer to as 'the Zimmerman Argument', or 'ZA' for short.

Generally speaking, ZA is an argument against ontological pluralism (IMO2), the view that says there are many, many different kinds of compound material objects, including the garden-variety material objects of our everyday experience such as artefacts and living organisms. According to the conclusion of ZA, the only kinds of compound material object that have instances are those kinds that are usually thought to constitute living objects and artefacts, e.g. cellular tissue, water, clay, etc. To put it another way, the only compound material objects that exist are stuffs, or *masses* according to ZA. Before laying out the premises of ZA, I need to explain some distinctions having to do with *a theory of masses*, since the plausibility of one theory of masses over others lies at the heart of ZA.

What are masses or stuffs? First of all, masses (or stuffs) are the referents of mass terms. It is reasonable to make a syntactic distinction between *mass terms* and *count nouns*. Mass terms differ from count nouns in the following ways. Syntactically, count nouns (a) admit of pluralization, e.g. 'the *trees* over there are about to lose their leaves', (b) can occur with numerals, e.g. 'those two *trees* are my favourite ones', (c) take 'a' and 'every' in the singular and 'few' and 'many' in the plural, e.g. 'every *tree* loses a few *leaves* every hour in the fall'.[27] In contrast to count nouns, mass terms are nouns that always take singular verbs, cannot occur with numerals, and take articles such as 'much' and 'little'. For example, consider the ways that 'earth', 'water' and 'time' do and do not function syntactically in the following phrases and locutions: 'the earth is wet' (mass nouns such as 'earth' take singular verbs); 'two earth' and 'single earth' (mass nouns do not occur with numerals); 'much water is required if the crops are to grow properly' and 'little time is required for the project' (mass terms take articles such as 'little' and 'much').

Distinguishing mass terms from count nouns in the ways I have been doing puts words such as 'gold', 'water', 'time' and 'freedom' in the 'same *syntactic* category' (Zimmerman 1995, p. 53). However, the referents of 'gold' and 'freedom' clearly belong to different *ontological* categories. Thus, it is important also to distinguish between *abstract* mass terms and *concrete* mass terms. Zimmerman suggests that it is only the latter that requires development in any ontology of *material* objects (1995, p. 53). Following Tyler Burge,[28] Zimmerman defines a 'concrete mass term' as follows:

(CMT) 'K' is a concrete mass term = (df) 'K' satisfies the syntactic criteria for mass terms, and 'Necessarily, any sum of parts that are K is K' is true. (1995, p. 53)[29]

Zimmerman assumes that the referents of mass expressions – masses – are the entities that constitute all material objects: if x constitutes y, then x is a mass. To see why he thinks this is so, take a standard example of material constitution: a statue and the clay that constitutes it. The term 'the clay' is a concrete mass term. Unlike a count noun, it cannot by pluralized, and does not occur with numerals.[30] The clay that constitutes the statue is therefore rightly understood to be a *mass* of clay, or some clay.

However, someone might object that the sentence 'the clay constitutes the statue' should be read, 'a piece of clay constitutes the statue'. Indeed, as we have seen, Lynne Rudder Baker thinks that a statue is not constituted by a mass, but rather by a *piece* of some stuff-kind (recall that pieces but not portions (masses) of marble can survive changes with respect to their parts). Perhaps some constituting entities are not masses after all but really just pieces of matter (note that 'a *piece* of marble' is a count noun and not a mass expression). As we shall see, however, Zimmerman's claim that all constituting objects are masses is not really controversial. This is because there are different theories of masses, that is, different theories about the nature of the referents of mass expressions. Thus, the first premise of ZA reads:

(1)   The referents of mass terms are called 'masses' and, if x constitutes y, then x is a mass.

So far we have simply stipulated that constituting entities are masses. What exactly are masses? Zimmerman suggests that there are two basic approaches to masses, each one of which he calls a 'theory of masses' (1995).[31] Zimmerman notes that the following theories of masses have been defended in the recent literature: (a) masses are individual material objects; (b) masses are sets that have individual material objects as their members, or alternatively, masses are a plurality of individual material objects.[32] Let us take a look at both of these theories of masses in turn.

Zimmerman himself favours mass theory (a), and he refers to it as 'the sum theory of masses' (1995, p. 55). The sum theory of masses has it that 'each referent of an expression like "the K" or "[some] K" is the mereological sum of one or more bits of K' (1995, p. 55). Furthermore, every referent of 'the K' or 'some K' is itself a part of a mereological sum of the world's total K. Suppose that the stuff that interests us is clay, and we have before us three statues, all of which are constituted of some clay (call the masses of clay that constitute the statues 'a', 'b' and 'c', respectively). The sum theory of masses tells us that a, b and c are proper parts of a mass of clay that is an individual material object identical to all of the clay currently existing in the world. In addition, a, b and c each have proper parts that are masses of clay. The mereological sum of mass a's proper parts (all of which are masses of clay) is identical to a (of course, the same goes for b and c). Furthermore, according to the sum theory of masses, the mereological sum of a, b and c is identical to an individual material object (the mereological sum of the clay in all three statues) – call it 'x'. Imagine that someone smashes a, b and c together (suppose that the clay in the statues is still wet). According to the sum theory of masses, doing so would not bring anything new into existence, but this is not for the reason that one might first suppose. As long as a, b and c exist, x exists according to the sum theory, no matter what the spatial relationships of a, b and c. The only way to cause x to come into or go out of existence on the sum theory of masses is for a, b or c to come into or go out of existence. Thus, the sum theory of masses is committed to the reality of scattered objects.[33] As strange as this implication of the sum theory may seem, Zimmerman points out that it is the majority view of philosophers and linguists working on mass terms and masses (1995, p. 58). However, as ZA progresses, we shall see that there are good reasons for accepting the sum theory of masses.

Finally, we should note also that Zimmerman thinks that the K one may have in mind in speaking about a mass is *matter itself* as in the phrase 'my body is constituted of the same matter today as it was yesterday'. Thus, although my body is composed of some blood, flesh, bone, etc., the mereological sum of these masses of different kinds is itself an individual material object according to the sum theory of masses, namely, a mass of cellular tissue, or alternatively, a mass of matter (1995, p. 77).

A second way to approach the nature of masses is to treat them as pluralities.[34] According to the plurality theory of masses, mass terms don't refer to individual material objects, but rather always refer to a number of individual material objects. There are such things in reality as clay *atoms*, that is, bits of clay that do not themselves have bits of clay as proper parts. (Divide a clay *atom* and what results are masses that do not belong to the kind *clay*.) On the plurality theory of masses, 'the clay' in the sentence 'the clay (and nothing but the clay) constitutes the statue' refers to a plurality of individual material objects – a plurality of clay atoms – and not one material object made of clay.[35]

Consider the following reason for treating a constituting mass as a plurality of individual objects instead of as a single material object as does the sum theory of masses. Assume for the sake of argument that a constituted object, e.g. a statue, is itself an individual material object, or whole (assume also that there are no temporal parts and that, necessarily, identity is transitive). According to the *sum theory* of masses, the mass of clay that constitutes the statue is itself an individual material object, or whole. On such a set of assumptions, there will be spatially coincident objects. But according to the *plurality theory* of masses, the clay that constitutes a statue is a plurality of objects, i.e. a number of clay atoms, and so the clay that constitutes a statue is not an individual material object. Since the clay and the statue are not two individual material objects existing in the same place at the same time on the plurality theory, there remains no worry about spatially coincident material objects. The plurality theory appears to preserve more of our common-sense intuitions about material objects than the sum theory of masses.

Given these two different theories of masses, we can formulate the second premise of ZA:

(2)   A mass of K is either (a) an individual material object x that (i) is identical to the sum of its parts, where x's parts are themselves masses of K, (ii) is itself a part of a material object that is identical to the sum of all the world's K, and (iii) is possibly a part of a material object y, where y is identical to the sum of those masses of K and *not-K* that are physically contiguous with one another, or (b) a plurality of material objects, each one of which is some K.[36]

The next premise of ZA distinguishes two general approaches to thinking about the relation of material constitution. Zimmerman proposes that these two general approaches to material constitution take their shape from two different ways of seeing the natures of the *relata* of the constitution relation. A *single category* approach to material constitution treats both the constituting mass and the constituted thing as belonging to 'the same basic ontological category' (1995, p. 70). In contrast, a *multiple category* approach to material constitution has it that the constituting mass and the constituted thing belong to different ontological categories; for example, the constituting mass is construed as a *plurality* of material objects and the constituted thing as an individual material object. So the third premise of ZA reads:

(3)   There are two general approaches to understanding the constitution relation: (a) a single-category approach, in which case both of the *relata* of the constitution relation belong to the same ontological category, and (b) a multiple-category approach, in which case the *relata* of the constitution relation belong to different ontological categories.

Zimmerman argues that there are three possible *single*-category approaches to material constitution (1995, pp. 70–1). One single-category approach has it that both the constituting mass and the constituted thing are individual material objects that are distinguished from one another on the basis of having different persistence conditions. Thus, the mass of clay and the statue are each individual material objects, or material wholes; they are two spatially coincident material objects. This is, of course, the standard account of material constitution. According to the options Zimmerman has offered with respect to theories of masses, the standard account would seem to be committed to the sum theory of masses: masses of K are *individual* material objects that are mereological sums of masses of K as well as part of the total sum of the world's K.

The metaphysician who admits that material objects have temporal parts takes a second sort of single-category approach to material constitution. Recall that a temporal part is a temporal atom of a physical object, and a physical object is assumed to be identical to the mereological sum of these temporal parts by a temporal-parts theorist. For the temporal-parts advocate, the constituting mass of clay has a number of temporal parts, only some of which are identical to the temporal parts of the statue. The temporal-parts advocate thus attempts to preserve the difference between the statue and the clay – they have different sets of temporal parts – without admitting coincident objects. Notice that by this approach the advocate of temporal parts also treats the statue and the clay as entities that both belong to the same ontological category, namely the category of individual material object.

The purveyors of 'a kind of weakened, pseudo-identity relation' (1995, p. 71) round out the single-category approaches to material constitution. According to this single-category approach to material constitution, the constituting mass and the constituted object are both individual material objects that are indeed identical, but only contingently, temporarily or according to one sortal but not others. Given these three possible single-category approaches to material constitution, we have premise (4) of ZA:

(4)   There are three single-category approaches to material constitution: (a) the constituting mass and the constituted thing are both individual material objects that share the same physical properties but differ in their persistence conditions; (b) the constituting mass and the constituted thing are both individual material objects that share some of the same temporal parts, and (c) the constituting mass and the constituted thing are both individual material objects that are identical relative to sortal, time or possible world.

Now let us consider three multiple-category approaches to material constitution.[37] The first multiple-category approach to material constitution has it

that constituting masses, e.g. some clay or some cellular tissue, *are not* individual material objects, whereas constituted objects, e.g. artefacts and living
organisms, *are* individual material objects. This is the approach to material
constitution taken up by the advocate of the plurality theory of masses. The
advantage of this multiple-category approach to material constitution is
that it preserves our intuition that constituted things such as toys, trees and
tigers are all individual physical objects, without having to admit spatially
coincident objects, temporal parts or some eccentric interpretation of the
identity relation.

The remaining multiple-category approaches are either reductivist or eliminativist in orientation: constituting masses *are* individual material objects,
whereas constituted objects turn out not to be material objects after all. On a
reductivist approach to constituted objects, constituted objects (or certain kinds
of constituted object) are not in reality *material* objects, but are rather objects
that belong to some different ontological category. For example, according
to one reductivist approach to material constitution, a constituted object can
be thought of as being identical to an event or process that 'passes through'
different constituting masses at different times, in the way that a wave passes
through different masses of water at different times (1995, pp. 71–2). In contrast to a reductivist approach to constituted objects – which admits that
constituted objects exist – the *eliminativist* approach to constituted objects (or
certain kinds of constituted object) has it that such objects do not exist at all.
According to the eliminativist, constituted objects may appear to exist – if
only for the reason that we commonly talk about them as though they
exist – but in reality they are non-existent. Zimmerman argues that mereological essentialists (those who deny IMO4) are eliminativists about constituted
objects (1995, p. 71). This is because a constituted object such as a human
body is thought to be the kind of object that can survive part replacement
whereas mereological essentialists believe that all material objects are essentially related to their parts. So premise (5) of ZA lays out the different sorts of
multiple-category approaches to material constitution:

(5) There are three multiple-category approaches to material constitu-
     tion: (a) constituted objects are individual material objects, whereas
     constituting masses are sets or pluralities; (b) constituted objects are
     in reality non-material entities, e.g. processes or events, whereas con-
     stituting masses are individual material objects; (c) constituted objects
     are mere virtual objects, whereas constituting masses are individual
     material objects.

The remaining premises of ZA put forward answers to questions having to
do with the constitution of masses themselves. Are there ultimate mass kinds or

is every mass itself constituted by another kind of mass? If there are ultimate masses, are they atomic or non-atomic in nature?

Let us first take up the question whether there are ultimate mass-kinds – kinds whose instances do not have parts that differ from the wholes to which they belong in kind. In order to answer this question, it will be useful to have definitions of 'homeomerous mass-kind' and 'heteromerous mass-kind':

(HO) K is a homeomerous mass-kind if and only if every mass x of K is such that, for any part p of x, p is some K.[38]

(HE) K is a heteromerous mass-kind if and only if every mass x of K is such that at least one of x's proper parts does not belong to the same mass-kind as x.[39]

As Zimmerman points out, the distinction between homeomerous and heteromerous mass-kinds has some pre-theoretic plausibility (1995, p. 62). There are obviously instances of heteromerous mass-kinds in the world, e.g. every mass of clay is composed of masses that are not some clay but, for example, are masses of H_2O. But are there instances of homeomerous mass-kinds, kinds whose instances are K, as Zimmerman puts it, 'through-and-through' (1997, p. 19). As Zimmerman points out, the only alternative to such a view is to admit what he calls an 'Oriental Boxes' theory of matter (1995, p. 76). This is the view that states that for any mass x of K, x is constituted by a mass that is not-K. Or to put it another way, the proposal under consideration is that all stuff-kinds are in fact heteromerous. In that case, every material object would be constituted of stuff that is itself constituted by a different kind of stuff, which is itself constituted by a different kind of stuff, *ad infinitum*.

So are there ultimate mass-kinds, masses that are homeomerous and not heteromerous? Zimmerman wants to suggest that, at the very least, the Oriental Boxes theory of matter is not necessarily true (1995, p. 76). There is a possible world in which there are in fact homeomerous masses. However, Zimmerman also confesses to wondering 'whether there isn't some hidden impossibility' with respect to the Oriental Boxes theory of matter (1995, p. 76). Whatever may be the case for the possibility of the Oriental Boxes theory of matter, ZA assumes that there are in fact, ultimate, homeomerous masses.

(6)   There are instances of ultimate mass-kinds, that is, there exist masses of some mass-kind K that are homeomerous.

The next premise of ZA addresses the question whether homeomerous masses are atomic or non-atomic in nature. As Zimmerman writes, homeomerous masses 'must either terminate in simple (i.e. [proper] partless) bits of K or be infinitely divisible into ever smaller bits of K' (1997, p. 20). The former view is that of the atomist whereas the latter view is that of the non-atomist.

First of all, I need to spell out Zimmerman's notion of a 'K-atom', which is a notion, the reader should note, that is applicable to the positions of both the atomist and the non-atomist:

(KA) x is a K-atom if and only if 'x is a mass of K, but no proper part of x is a mass of K' (1995, p. 64).

According to KA, it is possible for x to belong to a heteromerous kind. A water molecule is a heteromerous mass (it has proper parts that do not belong to the stuff-kind 'water'), and yet it nevertheless is an example of a water *atom* according to KA. This is because none of the proper parts of the mass of water that is identical to one molecule of H_2O are themselves masses of water. In applying KA to cases involving homeomerous masses, it appears as though one is committed to the view that such K-atoms are material simples, i.e. material objects having no *proper* parts. To see this, check KA and HO. By HO, if x is a homeomerous mass K, then every part of x – whether an improper or a proper part of x – is some K. But by KA no proper part of x is a mass of K. Indeed, since no part of x is not-K and x is a K-atom, this means that x has only one part, namely x itself, or x's improper part. But a mass of K that has no proper parts is a material simple. Therefore, a K-atom of a homeomerous stuff-kind is a material simple. With the notion of a K-atom in hand, let us see what follows for the two different approaches to masses (the sum theory and the plurality theory) when they are given an *atomistic* interpretation.

An atomist thinks that the ultimate constituents of material reality are material simples, i.e. entities that have no proper parts. If an atomist favours the sum theory of masses, then every mass of K is identical to a mereological sum of K-atoms. In fact, even K-atoms themselves would count as masses of K according to the sum theory (1995, p. 64). For example, assuming for the sake of argument that water is a homeomerous stuff-kind, according to the sum theory of masses, a visible mass of water would be identical to a particular mereological sum of water atoms. In addition, the water atoms that form the parts of the visible mass of water would themselves count as masses of water. The visible mass of water in turn is itself a proper part of at least one individual material object, e.g. the individual material object that is the mereological sum of all the water in the world.

Now suppose the atomist takes up the plurality theory of masses. Since, according to the plurality theory of masses, mass terms always refer to a plurality of individual physical objects and not to a single physical object, the ultimate constituents of a mass cannot themselves be masses on the plurality approach. Thus, according to the plurality theory, although each K-atom must be thought of as being some K, it cannot be a mass of K. Thus, we have to make the following modification of KA if we are to employ it in tandem with the plurality theory of masses:

(KA*) x is a K-atom if and only if 'x is some K, but no proper part of x is some K' (1995, p. 95).

If x is a mass that belongs to a homeomerous mass-kind K, and atomism is true, according to the plurality theory of masses, x is identical to a plurality of K-atoms, each of which is some K but not a mass of K. Assuming atomism is true, K-atoms are not themselves masses for the plurality theory of masses because masses are by definition pluralities of objects and each K-atom is a proper partless individual object.

But suppose atomism is false. Suppose that instead of there being such things as homeomerous K-atoms, it is rather the case that every instance of a homeomerous mass-kind K is infinitely divisible into some K. In that case, how should one speak about instances of homeomerous mass-kinds? We might speak about non-atomic homeomerous masses as follows:

(NAHM) x is an instance of a homeomerous mass-kind K if and only if x is a mass and x has proper parts, x is such that every one of x's proper parts is a mass of K, and every proper part of x has a proper part that is a mass of K.

As we have seen, according to the sum theory of masses, the proper parts of a mass x of a homeomerous mass-kind K (call them 'the ys') are individual material objects – assuming x has proper parts, of course. Furthermore, the ys are all masses of K and x is a proper part of at least one individual material object, that is the mereological sum of all the world's K. Now in a world where non-atomism is true, the sum theory of masses says that the ys themselves have proper parts (call them the zs), each one of which is itself a mass of K. Of course, the zs also have proper parts that are masses of K, *ad infinitum*. The sum theory of masses has no problem making sense of the possibility that there are no K-atoms for any homeomerous mass-kind K.

However, the plurality approach to masses does not fair so well on the supposition that atomism is false. As Zimmerman argues,

The sort of 'mere plurality' picked out by a plural referring term is not a *single thing* of any sort – that is just the difference, I take it, between the denotations of plural referring terms on the one hand, and sums and sets on the other. So to identify masses with mere pluralities that were pluralities of pluralities 'all the way down' would be to identify masses with *nothing at all*. (1995, pp. 99–100)

If one supposes that a homeomerous mass is atom-less, as far as the plurality approach to masses is concerned, one would never get to the referents of locutions such as 'some K' and 'the K', since *every* mass of K is divisible into smaller *masses* of *K*. Imagine that water is a homeomerous stuff-kind. According to the plurality approach to masses, 'a mass of water' refers to a plurality of individual physical objects (call them 'the *xs*'). Now because we are supposing that water is homeomerous and atom-less, the *xs* are themselves masses of water. That means that in order for the original locution 'a mass of water' to refer to something, the *xs*, which are themselves masses of water, must each refer to a plurality of individual physical objects. But, according to non-atomism, the proper parts of the *xs* all turn out to be masses of water as well; in addition, the proper parts of the proper parts of the *xs* will also turn out to be masses. Thus, an infinite regress ensues, with no hope of ever arriving at a real plurality of individual material objects. Hence, the plurality approach to masses is incompatible with the existence of atom-less, homeomerous stuffs.

Given what we have said about ultimate stuff-kinds and atomism, we can record the following additional premises of ZA:

(7) For any mass *x* of a homeomerous mass-kind *K*, either (a) *x* is ultimately constituted by *K*-atoms, which themselves are some *K* but have no proper parts, or (b) *x* is infinitely divisible into ever smaller masses of *K*.

(8) The plurality theory of masses is incompatible with non-atomism, i.e. the existence of atom-less, homeomerous masses.

(9) The sum theory of masses is compatible with both atomism and non-atomism.

Next, ZA assumes for the sake of argument that there are no such things as temporal parts, i.e. IMO3 is true, and that identity is not relative to sortal, possible world, or time, i.e. IMO6 is true (1995, p. 86). Thus,

(10) It is not the case that individual material objects have temporal parts or identity is relative to sortal, possible world, or time [∼(4b) and ∼(4c)].

Zimmerman further argues in some detail that coincident objects are not to be countenanced (1995, pp. 86–90). Of course, it does not follow from the supposition that the constituting object and the constituted object are not spatially coincident material objects that the constituting and the constituted objects are identical. Now Zimmerman does accept the soundness of the standard argument for the non-identity of the constituting mass and the constituted object: a constituting mass cannot be identical to the object it constitutes since the latter but not the former can survive replacement of its parts (1995, pp. 70ff.). Nevertheless, not every account of material constitution is committed to the spatial coincidence of two individual *material* objects. For example, none of the multiple-category accounts of material constitution entail the possibility of spatially coincident material objects (1995, p. 91). This is because multiple-category approaches to material constitution have it that either the constituted object or the mass that constitutes it is not an individual material object. According to a multiple-category approach the constituting mass and constituted thing are indeed spatially coincident, but in a way analogous to how a concrete individual, e.g. a house, and one of its properties, e.g. *being pink*, are spatially coincident – surely, a non-objectionable sort of spatial coincidence. At any rate, according to ZA:

> (11)   Coincident individual material objects are not to be countenanced [∼(4a)].

Given the premises above, a few of the proposed approaches to material constitution are no longer available to us. First, from 4, 10 and 11 it follows that

> (12)   It is not the case that a single-category approach to material constitution is correct [∼(3a)].

In addition, from premises 3 and 12 we can conclude:

> (13)   A multiple-category approach to material constitution is correct [(3b)].

The last controversial premise in ZA involves the question whether or not we should take non-atomism seriously. Zimmerman strongly suggests that we should, and for a number of reasons. First of all, non-atomism could be true. There is a possible world in which there are homeomerous masses of K, and no masses of K are K-atoms. Furthermore, it is possible that non-atomism is true in the actual world. It is true enough that our current physical theories suggest that fundamental particles are simples, and not atom-less masses. However, it hardly needs mentioning that scientific theories (even the most entrenched) are often times revised, and sometimes overturned. Perhaps the fundamental physical theories of the future will take a turn away from atomism. Lastly, there may be good *a priori* arguments for the truth of non-atomism – how could extended physical objects be composed of non-extended, proper partless objects?[40] It would be useful to have a theory of masses and a theory of material constitution that accommodated the possibility that atomism is false. Given these reasons for taking seriously the possibility that atomism is false – and in order to keep things simpler – my formulation of ZA will assume

> (14)   Atomism is false.[41]

As I pointed out above, the plurality theory of masses cannot accommodate homeomerous non-atomistic masses. However, we also saw that the sum theory of masses can accommodate the possibility that either atomism or non-atomism is true. Given premise (14), coupled with the fact that the plurality approach to masses is incompatible with the truth of non-atomism 8, it follows that:

(15)   A plurality approach to masses is false [∼(2b)].

In addition, from premises (2), (9) and (15) it follows that:

(16)   The sum theory of masses is correct [(2b)].

From (15), it also follows that:

(17)   It is not the case that constituted objects are material physical objects constituted by masses construed as pluralities [∼(5a)].

But the following premise is also true:

(18)   Multiple-category approaches to material constitution (5b) and (5c) are both compatible with the falsity of atomism.

Therefore, given premises (5), (13), (16), (17) and (18), the conclusion of ZA reads:

(19)   Therefore, either (a) constituted objects are in reality non-material entities, e.g. processes or events, and constituting masses are individual material objects, or (b) constituted objects are mere virtual objects, whereas constituting masses are individual material objects [(5b) or (5c)].

Recall that constituted objects are the referents of count nouns such as 'toy', 'tree', and 'tiger'. Thus, if ZA is a sound argument, then there are not as many compound material objects in the world as we would typically suppose. Indeed, the only true material objects according to ZA are the masses that constitute things such as toys, trees and tigers; toys, trees and tigers are not themselves individual material objects. Why? First of all, because we have assumed the truth of IMO3 (material objects don't have temporal parts), IMO5 (spatially coincident material objects are not to be countenanced) and IMO6 (necessarily, identity is transitive). Secondly, the sum theory of masses is correct. But the sum theory of masses entails that, for any artefact or living organism, there is an individual material object that is spatially coincident with it, a mass of matter that is not identical to that artefact or living organism. This leaves us with the following two meagre possibilities: either it is the case that artefacts and living organisms are events akin to waves passing through masses of water, or else they are simply virtual objects. Thus, ZA can be seen as a serious challenge to the truth of ontological pluralism (IMO2).

I have examined in detail the views of three contemporary philosophers on the nature of compound material objects and related these respective views to the answers they give (or at least, would give) to the PMC. All three of these philosophers have it that there are compound material objects, and that these objects endure, rather than perdure, through time. Furthermore, they all agree that talk of the identity of material objects should be conducted within the context of a traditional approach to the identity relation. Interestingly, all three of these philosophers also assume that constituting masses – however they are construed – are entities that continue actually to exist in the objects they constitute just as they did before they constituted anything. This is a key metaphysical assumption and one that, as we shall see, Thomas Aquinas rejects.

With the theories and arguments of Baker, Van Inwagen and Zimmerman in mind, I now turn to developing in several stages Thomas Aquinas's views on compound material objects. I begin in Chapter 3 with a discussion of Aquinas's concept of substance, a concept integral to any discussion of Aquinas's metaphysic of material objects.

Notes

1. See: Baker, L. R. (1997) 'Why constitution is not identity', *The Journal of Philosophy* 94: 599–621; Baker, L. R. (1999a) 'Unity without identity: a new look at material constitution' *Midwest Studies in Philosophy* 23: 144–65; and Baker 2000.
2. Baker 1997.
3. Baker 2000, p. 29.
4. Baker 2000, p. 28.
5. This passage brings out a view of Baker's that is fairly idiosyncratic in the constitution literature, namely, that constitution is an irreflexive, asymmetric and non-transitive relation (2000, pp. 44–5). According to Baker, if x constitutes y, then y does not constitute x (2000, p. 33). As Baker puts it, thinking of constitution in this asymmetrical fashion, 'induces a kind of ontological hierarchy – a hierarchy that reverses the usual reductive hierarchy' (2000, p. 33). Cf. Rea 1997, p. lii and Zimmerman 1995, p. 74.
6. See, e.g., Burke 1994a and Rea 2000.
7. Baker notes that there are four classes of (non-ordinary) properties that cannot be had derivatively, but are only ever had non-derivatively. See Baker 2000, pp. 48–9 for discussion.

8. See, e.g., Wiggins 1968.

9. Cf. Burke 1997 and Zimmerman 1995, and see my discussion of Zimmerman's views in the sequel.

10. See also Baker 2000, pp. 33–4 n. 19.

11. However, this claim comes on the heels of her comment, 'I use the term "aggregate" to singularize a plurality of things'. This makes it tempting to think that Baker is simply making a grammatical move in saying that an aggregate is an individual object. But for the reasons I spell out above, this cannot be what Baker has in mind.

12. The standard example in the contemporary philosophical literature of a case of material constitution: a statue and the clay that constitutes it (see, e.g., Burke 1992). One reason for thinking that the statue and the clay are not identical is that they have different essential properties. Although the statue can survive the loss of some of its parts, it cannot survive a radical change in shape, e.g. being melted. Thus, a particular statue does not have *being made of just these parts* as an essential property, but it does have *being such that its shape resembles that of a human being* as an essential property. In contrast to the statue, the portion of clay *can* survive a radical change in shape, but it cannot survive the loss of any of its parts. Thus, the clay does not have shape as an essential property, but it does have *being made of just these parts* as an essential property. Note that this understanding of 'portion' as an object essentially related to its proper parts is identical to the bonding account of an aggregate: an object whose proper parts (the *x*s) are bonded to one another and whose identity is contingent upon the *x*s remaining bonded to one another.

13. See also condition (a) of Baker's definition of 'constitution' (2000, p. 43).

14. See, for example, Van Inwagen 1990a and 1981 (page citations of this work come from the reprint in: Van Inwagen (2001) *Ontology, Identity, and Modality: Essays in Metaphysics*, Cambridge: Cambridge University Press). See also: (1997a) 'Foreword' in M. Rea (ed.) *Material Constitution: A Reader*, Lanham: Rowman & Littlefield, pp. ix–xii.

15. See, for example: Van Inwagen 2001, pp. 79ff.; 1990a, pp. 3–4 and (1990b), 'Four-dimensional objects', *Nous* 24: 245–55. For some of Van Inwagen's reservations about IMO6, see 1990a, p. 3.

16. Van Inwagen 1990a.

17. Van Inwagen labels this question about composition 'Special' since he distinguishes it from another question that concerns composition that he calls the 'General Composition Question'. The General Composition Question has to do with offering a non-mereological analysis of the term 'composition' (see Van Inwagen 1990a, pp. 38–51 for discussion). Given my purposes, I will not have occasion to say anything more about the General Composition Question.

18. Van Inwagen also has an interesting discussion of atomism and non-atomism in: (1993), *Metaphysics* Boulder, CO: Westview Press, pp. 22–8.

19. Van Inwagen in a number of places contrasts his own atomistic views with those of Aristotle on this score. For example, Van Inwagen notes, 'one might suppose – it can be argued that this is Aristotle's view of the matter – that organisms have

no proper parts, that they are entirely composed of absolutely continuous stuffs'
(1990a, p. 98). Of course, he assumes that this position is empirically false. Nevertheless, Van Inwagen claims that Aristotle's theory of material objects – or at
least a certain interpretation of Aristotle – is the one closest (among notable theories) to his own (1990a, pp. 15, 92).

20. An example of two objects being fastened might be two blocks of wood joined in
     virtue of some wood screws.

21. An example of cohesion between material objects might be two blocks of wood
     joined by an industrial-strength glue.

22. An example of fusion: the ingredients for bread, after they have been mixed
     together. See Van Inwagen 1990a, p. 59 for a more scientifically astute example.

23. By 'right-hand side of the answer' I mean (with Van Inwagen) that part of the
     answer that comes immediately after 'if and only if'.

24. Van Inwagen 1990a, 89. Van Inwagen does discuss the possibility (in fact, the
     actuality) of lives overlapping. He suggests that the only possible scenario where
     this occurs is in a living organism composed of cells. In this case alone, the activity
     of some cells (call them 'the *x*s') constitutes a life and the activity of some physical
     simples constitutes the lives of the *x*s. However, Van Inwagen thinks that there is
     clearly a relationship of subordination involved in this case, with the lives of cells
     being subordinate to the life of the organism that those cells compose (1990a, 89).
     Talk of 'subordination' suggests that Van Inwagen thinks that cases involving the
     subordination of one life to another do not in fact offer us counter-examples to
     the view that lives are characteristically jealous. And this view of Van Inwagen's
     seems plausible. Indeed, a living organism's being composed of living cells does
     not provide a case where two lives are sharing precisely the same space.

25. For additional arguments in defence of Van Inwagen's position on the SCQ, see
     1990a, pp. 98–141.

26. See, e.g., Zimmerman 1995; (1996) 'Could extended objects be made out of
     simple parts? An argument for "atomless gunk"', *Philosophy and Phenomenological
     Research* 56: 1–29, and (1997) 'Coincident objects: could a "stuff ontology" help?',
     Analysis 57: 19–27.

27. Zimmerman 1995, p. 53.

28. Burge, T. (1972) 'Truth and mass terms', *Journal of Philosophy* 69: 263–82.

29. I follow Zimmerman in using '*K*' as a 'schematic letter replaceable only by mass
     terms' (1995, p. 54).

30. Zimmerman notes a context in which the term 'clay' might be pluralized: when
     one is speaking of different kinds of clay. For example, an artist might have occa-
     sion to say 'I used several different clays in the process of making this statue'.
     As Zimmerman points out, this 'kind of' reading of a mass term is not what is at
     issue in the present context. What is at issue here is the nature of *particular* material
     objects (1995, p. 54 n. 4).

31. Zimmerman also notes a third theory of masses in the recent literature where
     masses belong to 'an ontological category located somewhere between mere plur-
     alities and individual physical objects, or wholes' (Zimmerman 1997, 20).
     Because the theory has received limited treatment in the literature, I don't say

anything else about it here. For exposition of this third view of masses, see Burke 1997. For a critique of Burke's view, see Zimmerman 1997.

32. Following Zimmerman (1997, pp. 19ff.), I am using the term 'plurality' as shorthand for the referent of a plural referring expression such as 'the xs', and since plural referring expressions do not necessarily have single objects as their referents, neither do the referents of 'plurality'.

33. A scattered object is a physical object x at least one of whose parts is separated from x's other parts spatially or by a physical object that is not a part of x.

34. Or, alternatively, as sets. I choose to focus on masses construed as pluralities rather than as sets to simplify matters here. Furthermore, nothing significant hangs on the difference between treating masses as sets or pluralities for my purposes.

35. Thus, 'atom' in this context is being used in a technical sense to refer to the smallest bit of a stuff-kind K. See the sequel for more discussion of this understanding of 'atom'.

36. As Zimmerman points out, it is possible to develop a hybrid theory of masses, where, for example, some masses are pluralities and others are individual material objects, or wholes (1995, p. 60). Zimmerman never develops this idea, and for the purposes of this chapter neither will I. However, see Chapter 7 for some discussion of the possibility that Thomas Aquinas's views on material objects can be understood along these lines.

37. See Zimmerman 1995, pp. 71–2; 1997, p. 20.

38. Recall that 'a part of x' may refer to either x's improper part, that is, the part of x that is identical to x, or a proper part of x, a real part of x that is not an improper part of x.

39. Cf. Zimmerman 1995, p. 62 where Zimmerman sets out more precise definitions of homeomerous and heteromerous masses.

40. See Zimmerman 1995, pp. 97–8. For an extensive treatment of *a priori* arguments for non-atomism, see Zimmerman 1996.

41. Zimmerman's own argument moves from the possibility of the falsity of atomism to the conclusion that theories of masses incompatible with such a possibility should be rejected. See 1995, pp. 105–6 and 1997, p. 20 n. 6.

3

Aquinas on Material Substances

Thomas Aquinas holds the view that there are many, many kinds of material objects in the world. Some of these are material objects in a primary sense while some of them are not. And of the objects that Aquinas counts as material objects in the primary sense – objects that he calls 'substances' – some of these are simple and others compound. Furthermore, there are many different kinds of simple and compound substances for Aquinas. Aquinas thus subscribes to a metaphysic of material objects that I have been referring to as 'ontological pluralism'. In this chapter I explain one of the most important parts of Aquinas's metaphysic of material objects, namely his understanding of material substance. In order to explain Aquinas's views on material substance, I first say some things about his understanding of the concept of substance as well as those kinds of entities that don't count as substances for him. Second, I treat Aquinas's understanding of the nature of materiality and therefore discuss the sorts of features that all material substances have in common. Third, I discuss Aquinas's understanding of the extension of 'material substance'. Here I talk about the particular kinds of material objects that count as material substances for Aquinas, and say a few things about the things that don't.

Aquinas on the nature of substances

Aquinas's particular approach to the concept of substance requires that I make a few preliminary remarks. Aquinas nowhere gives a systematic presentation of his doctrine of substance, as contemporary philosophers who are working on such problems are accustomed to doing.[1] Aquinas's discussions of the nature of substance are usually woven into (very systematic) discussions of some theological topic. I attempt here to reconstruct, in a systematic way, Aquinas's views on the nature of substance. Rather than offer a chronological account of Aquinas's theory, my discussion proceeds from Aquinas's less finely tuned descriptions of substance to those that are more finely tuned.

Aquinas accepts from Aristotle the view that there are ten different 'categories' of being. These ten categories constitute ten different ways that things can *be*, although these ten ways of being do *not* share a common genus, i.e. the categories of being are not ten species of being, but instead are related in different ways to the 'primary' category of being, substance.[2] Thus, the context for Aquinas's discussion of substance (and material substance in particular) is the common-sense picture of the world captured by Aristotle's ten categories, one that has it that there is a distinction to be made between material objects on the one hand, and the attributes such objects possess on the other. The most fundamental difference in the categories is thus that between the category of substance – the primary mode of being – and the nine categories of accidental being. Aquinas's way of making the distinction between substance and accident is the focus of the first part of this chapter.

Given the tradition in logic that Aquinas inherits, he thinks that it is impossible to offer a proper definition of 'substance'. According to this logical tradition, every definition has two parts: a genus and a difference. But substance cannot have a genus since, if anything was the genus of substance, it would be being, and being cannot be a genus.[3] I will bracket this complication for the purposes of my discussion. As a defence of this procedure, let me point out that Aquinas himself sometimes explicitly brackets this complication, offering what he refers to as a 'definition' of substance.[4]

Let me make one last preliminary remark. Aquinas's discussion of the concept of substance is complicated by the fact that he recognizes that the concept of substance admits of several different senses.[5] Whereas 'substance' can also refer to the nature that is shared by individual things that belong to the same species, e.g. the humanity of Socrates and Plato, the sense of substance at issue for my purposes is the one that Aquinas refers to as 'primary substance',[6] or 'hypostasis',[7] that is a particular substance. I now turn to discussing this sense of substance for Aquinas.

Some incomplete definitions of 'substance'

Aquinas sometimes refers to a substance as a being in virtue of itself (*ens per se*).[8] This suggests the following preliminary definition of substance:

(S1) *x* is a substance $=_{df}$ *x* is a being *per se*.

Now given that Aristotle's ten categories form the primary backdrop for Aquinas's understanding of substance, we know accidents are not supposed to satisfy S1. Indeed, a particular accident such as *this instance of the colour scarlet* does not satisfy S1 since, although it clearly exists extra-mentally and so is in some sense a real being, a particular accident does not enjoy such an existence in virtue of itself (*per se*) but only in virtue of the substance that it modifies. So far, S1 looks pretty good.

But we run into problems with the locution '*per se*'. Now we might understand '*per se*' to mean 'through itself' or, more concretely, 'without dependence on another'. That would make it the case that, according to (S1), a substance is 'a particular being whose existence does not depend on that of another'. Now, although this has been a historically important definition of substance,[9] it is certainly not Aquinas's understanding of substance. For on such an interpretation of '*per se*' S1 entails for Aquinas that the only being that would count as a substance is God. This is because, for Aquinas, all beings other than God depend on God for their existence, whereas Aquinas thinks that God's existence is something self-explanatory and non-dependent.[10] Furthermore, one often finds Aquinas at pains to *exclude* God from the extension of 'substance'.[11] Aquinas certainly does not understand substance in accord with the definition, 'a particular being whose existence does not depend upon another'.

Certain passages in Aquinas's corpus suggest that S1 could be revised by taking 'existing *per se*' to mean more precisely 'existing not in another as a subject'. Thus, we would have the following definition of substance:

(S2) x is a substance $=_{df} x$ does not exist in another as a subject.[12]

S2 thus clearly relaxes the constraints on what can count as a substance, allowing there to be substances other than God. Note also that it allows many of the material objects with which we are familiar to count as substances, in contrast to S1. For even if S1 were modified so as to allow for there to be substances other than God, it still would not be clear that any *material* objects could satisfy S1. This is because material objects (usually) depend on other *material* objects for their continued existence, e.g. there is a sense in which all compound material objects depend for their existence on their parts, and most material objects also depend upon objects external to them in order to remain in existence, e.g. for nutritive purposes.[13] In contrast to S1, S2 makes clear the precise way in which a substance is non-dependent; one which (as we shall see) is compatible with a material substance's being dependent on things extrinsic to that substance for its existence. However, S2 fails to capture Aquinas's understanding of substance, since it still allows for God to count as a substance, a possibility Aquinas clearly rejects.

Having rejected S1 and S2 as satisfactory definitions of Aquinas's understanding of substance, we can note that Aquinas often offers something like the following definition of substance, one that makes use of S2's modifications of S1, while also adding a further modification:

(S3) x is a substance $=_{df} x$ has a quiddity to which it belongs not to exist in another as a subject.[14]

Aquinas thinks that God is an absolutely simple (non-compound) being. One thing this means for Aquinas is that God does not, properly speaking, *have* a quiddity, since God's quiddity (or form, or nature or essence) is identical to His being.[15] Therefore, God does not satisfy S3, and so S3 does not fail as a definition of substance on that score. As we shall see, Aquinas accepts that an object *x*'s having a quiddity to which it belongs not to exist in a subject is a necessary condition for *x*'s being a substance. But Aquinas sometimes offers a definition of substance that adds to S3 a further requirement for an object's counting as a substance. Before I explain this additional requirement, let me say a few things about beings that don't count as substances given the necessary condition for substancehood picked out by S3.

Aquinas on the nature of accidents

A definition of substance such as S3 is clearly designed to contrast substantial being from the being that accidents possess. Indeed, when Aquinas does offer a careful 'definition' of accident, he does so as follows:

> (A1) *x* is an accident $=_{df}$ *x* has a quiddity to which it belongs to exist in another as a subject.[16]

According to S3 and A1, the crucial difference between a substance and an accident is that an accident does whereas a substance does not exist in another as a subject. Aquinas would seem to mean that, if a being exists in another *as a subject*, then that being exists as a kind of *modification* of that in which it inheres. This may suggest that Thomistic accidents are on a par with the *properties* of contemporary metaphysics. Indeed, I think the comparison between Thomistic accidents and contemporary properties is a helpful one to make as long as one makes some important caveats. Like Thomistic accidents, the individual properties of contemporary metaphysics are entities that are somehow had or possessed by material (or immaterial) objects. However, there are some important differences between Aquinas's understanding of accident and the properties of contemporary metaphysics.

Contemporary philosophers often speak about *properties* as being either essential or non-essential.[17] In contrast, Aquinas distinguishes those *accidents* that are 'proper' (*propria accidenta*)[18] from those that are 'extraneous' (*extranea accidenta*).[19] Whereas the non-essential properties of contemporary philosophy are (or can be construed as) roughly equivalent to Aquinas's extraneous accidents, the same cannot be said for the essential properties of contemporary philosophy and Aquinas's proper accidents. One might offer the following definition of Aquinas's proper accident:

> (PA) *x* is a proper accident $=_{df}$ (1) *x* is an accident, (2) there is a species *F* such that *x* belongs to *F*, (3) if *y* is the subject in which *x* inheres, then

(i) y belongs to species A, and (ii) for any time t at which y exists, x inheres in y, and (iii) for any z such that z belongs to A, an accident belonging to F inheres in z.[20]

Aquinas's classic example of a proper accident is the feature of human beings, *being risible*, or *being able to laugh*.[21] By way of contrast to the Thomistic proper accident, contemporary philosophers define an essential property in roughly the following manner:

(EP) P is an essential property $=_{df}$ (1) P is a property and (2) for any object y that has P, y has P in every possible world in which y exists.[22]

Aside from the above definition construing properties as universals – Aquinas thinks of accidents as individual things in a way that is perhaps similar to the tropes of contemporary metaphysics – there is another key difference between Thomistic proper accidents and the essential properties of the contemporary metaphysician. For those contemporary philosophers who accept that objects have essential properties, Socrates might be described as having the following essential properties: (a) *being a human being*; (b) *being an animal* and (c) *being risible*. Aquinas accepts that all human beings have (a)–(c). But, by comparison, Aquinas thinks that only (c) counts as a proper accident of Socrates. He considers (a) and (b) to be features of Socrates that signify Socrates's essential being; they are not accidents of Socrates at all, even proper accidents.[23] This provides one reason for thinking Aquinas's understanding of accidents does not match up perfectly with the contemporary notion of a property.

With this difference between the contemporary notion of an essential property and Thomistic proper accident in mind, an extraneous accident might be defined as follows:

(EA) x is an extraneous accident $=_{df}$ (1) x is an accident, and (2) x is not a proper accident.[24]

Notice that an extraneous accident x *might* inhere in a substance y at every time in which y exists. For example, say x belongs to the species *having a dark skin colour*, and a human being, Fred, has x de facto for the duration of his life. Nonetheless, x is not a proper accident of Fred; some human beings do not have an accident belonging to the species *having a dark skin colour*. A proper accident is an accident that belongs to that substance because of what that substance is most fundamentally (i.e. because of its essence). Fred's having dark skin is not a function of what Fred is most fundamentally, i.e. a human being.

Before I return to my discussion of Aquinas's understanding of substance, I need to say something about another way that Aquinas uses the term 'accident'. 'Accident' in the sense that I spoke about above is sometimes referred to as a 'predicamental accident',[25] since such accidents are contrasted with beings belonging to the category or predicament of substance.[26] However, following Aristotle, Aquinas speaks not only of accidents in the predicamental sense, but also of *ens per accidens*, or accidental being.[27] By the term 'accidental being', Aquinas means to refer to a composition that has at least one predicamental accident as what we might call its 'form of the whole'. For example, consider the composition of Socrates and the accident *being virtuous*. Here Socrates plays the role of matter while *being virtuous* is the form of the composite virtuous-Socrates. *Being virtuous* is not a part of Socrates's essence (human being); nor is it a proper accident of Socrates. It is possible that Socrates exists without being virtuous, e.g. Socrates was not virtuous when he was three years old. As we shall see, Aquinas thinks that artefacts are accidental beings in this sense, that is composites of a substance and an extraneous (predicamental) accident.

In sum, predicamental accidents are beings that exist in a subject, that is, a substance, and are not part of the essence of that substance. We might say that such accidents 'modify' the substances to which they belong. Accidental beings by contrast are composites of a predicamental accident and some other being or beings, typically a substance or collection of substances.

Aquinas's understanding of substance

Recall that S3 has it that a substance is a being having a quiddity (nature) to which it belongs not to have existence in another as a subject. Aquinas offers a fourth and final definition of substance that can be seen as a modification of S3. This definition of substance shows up in Aquinas's discussions of the distinction between a substance and its parts – objects such as hands and souls.

Parts of substances come in at least two varieties for Aquinas. First, there are *integral* parts of a substance, i.e. physically extended quantitative portions of a whole that (usually) have some specific function with respect to the whole, e.g. the hands and feet of a human being. Second, there are the *metaphysical* parts of a substance, i.e. those parts of a substance that go some distance towards explaining the existence, identity and nature of that substance, i.e. that substance's form and matter.[28] It is specifically in the context of comparing the soul of a human being – one of a human being's metaphysical parts – with the integral parts of a human being that Aquinas offers us his most mature definition of substance.

For example, at ST Ia. q. 75, a. 2, Aquinas discusses the question whether the human soul is a *subsistent* entity. As the answer to the second objection makes clear, Aquinas equates what subsists with what exists *per se*. Since, as we have seen, existing *per se* (rightly understood) is a necessary feature of substantial being, the question at issue here might be whether the human soul *is a substance*. In the answer to the first objection of this article Aquinas makes a distinction between two ways in which something can be subsistent, or a '*hoc aliquid*' (literally, a 'this something'). A subsistent thing, or a *hoc aliquid* is either 1. something that subsists and is *incomplete* in its species, or (2) something that subsists and is *complete* in its species. Thus, it seems that any *hoc aliquid*, whether of the first or second variety, is something that subsists *per se*, or is something that exists *per se*. But not every *hoc aliquid* is complete in species. Getting clear on what Aquinas means by *being subsistent* and *being complete in species* will help us understand what he takes a substance to be since he thinks that only a *hoc aliquid* that has both of these features counts as a substance.

In order to get clear on what Aquinas means by an object's *being subsistent*, let us first recall that a subsistent thing exists *per se*. As I pointed out above, by something existing '*per se*' Aquinas does not mean to say that such a thing exists independently from other things *simpliciter* as much as he means to point out that that thing does not exist *in another as a subject* – some *x* that is existing *per se* is not a modification of some other being *y*. Therefore a subsistent being is not a modification of something else; rather, it is the sort of thing that is itself modified.

Aquinas's examples of non-subsistent things in ST Ia. q. 75, a. 2, ad1, give us further clues as to what he means by a *hoc aliquid* or a subsistent thing. Certain sorts of particular things cannot count as a *hoc aliquid* in any sense of the term, namely accidents and material forms.[29] Now accidents and material forms do exist as particulars.[30] But accidents and material forms are not subsistent entities since they always exist in another as a subject, although they do so in different ways.

In another place Aquinas further notes that what is subsistent, or exists *per se*, is always generated *per se*, while that which does not subsist (what exists in another as a subject) is always generated *per accidens*.[31] Accidents and material forms do not subsist because they only come to be or go out of existence insofar as the subsistent thing they modify comes to be or goes out of existence. Accidents are said to 'exist in another as a subject' for Aquinas because whenever accidents exist, they exist only as modifications *of* something that subsists, that is, something that is generated or corrupted.[32] Material forms are not accidents, but they are like accidents in that they do not come to exist or go out of existence except as the forms *of* the substances of which they are a part.

Aquinas's examples of non-subsistent things give some clarity to the notion of a subsistent thing: a subsistent thing is not a *modification* of something else, whereas accidents and (in a sense) material forms *are* modifications (of some subsistent entity). But in addition to a *hoc aliquid*'s being a subsistent thing, there is also the matter of whether or not a *hoc aliquid* is something complete in species. What does it mean to be complete or incomplete in a species?

Again, Aquinas's examples are helpful. At ST Ia. q. 75, a. 2, ad1, Aquinas mentions hands and human souls – both parts of a substance (albeit in different senses) for Aquinas – as examples of the kind of *hoc aliquid* that is *incomplete* in species. On the other hand a substance – such as a whole human being – is a prime example of a *hoc aliquid* that is complete in its species.[33]

But why are substances complete in their species, whereas parts of substances are not? A part of a substance is *what* it is in virtue of the whole to which it belongs whereas a substance is what it is in virtue of itself (*per se*). Aquinas makes a similar point in discussing a difference between substances and accidents. Unlike a substance, an accident is incomplete in its essence, since a thing has its essence in accord with the way it is defined, and one must include something extrinsic to an accident (i.e. a substance) in order to complete that accident's definition.[34] This makes sense given that Aquinas understands an accident to be a modification *of a substance*. Aquinas goes on to extend this idea of including something extrinsic to a thing in that thing's essence to the case of substantial forms as well. What is a substantial form? It is the form *of a substance*. Furthermore, it is something related to matter. Both the *substance* of which a substantial form *x* is the species conferring form and the *matter* that that substantial form actualizes are extrinsic to *x*, and yet the quiddity of that substantial form *x* includes a reference to both substance and matter.

Aquinas seems to have the same sort of idea in mind when contrasting a *hoc aliquid* that is incomplete in its species with a *hoc aliquid* complete in species. As in the cases of accidents and substantial forms, parts of substances are defined (they are *what* they are) by way of reference to something extrinsic to themselves, in this case, the substantial wholes to which such parts belong. A *human* hand is defined by way of reference to its parent substance, a human being. Thus, in contrast to a substance, whose essence or quiddity does not include a reference to something extrinsic to it, accidents, integral parts and metaphysical parts are all defined with reference to something extrinsic to them, and so they, and not substances, are said to be incomplete in their species.[35]

With these explanations of Aquinas's notions of subsistence and being complete in species, I can offer the following as Aquinas's most nuanced definition of substance:

(S4) *x* is a substance = $_{df}$ (a) *x* is a *hoc aliquid*, that is, a subsistent thing, and (b) *x* is complete in some species, that is *x* is able to be defined without reference to something extrinsic to *x*.

Although S4 makes use of terminology much different from S3, it nevertheless contains all the notes of S3 while adding to it a crucial modification. S3 has it that a substance has a quiddity to which it belongs not to exist in another as a subject. If some being x is a *hoc aliquid*, that means that x is both a particular and subsistent being. But a subsistent being is one that does not exist in another as a subject. Thus, just as S3 works to distinguish substances from accidents, so does clause (a) in S4. Furthermore, if x belongs to a species (whether completely or only insofar as it is a part of something that falls under a species), then x cannot be God, since God does not belong to a species or a genus.[36] Thus, just as S3 entails that God is not a substance – since God does not *have* his essence – so does clause (b) in S4. However, S4 offers a more complete account of substance than S3, since S4 distinguishes substances from the kinds of *hoc aliquid* that are incomplete in species, e.g. the human soul or the integral parts of material substances. Substances and not souls are complete in species; a substance is what it is in virtue of itself while a soul is what it is in virtue of the substance of which it is a (metaphysical) part.

I need to make one final remark about Aquinas's understanding of the concept of substance. Aquinas does not always use the term 'substance' in the technical sense that I have laid out, for example, in S4. He often uses the term loosely. For example, sometimes Aquinas refers to the human soul as a substance.[37] Furthermore, he sometimes refers to the integral parts of substances[38] and to artefacts[39] as substances. Now it is clear that Aquinas's considered opinion on souls, the integral parts of substances, and artefacts is that none of them count as substances, the reasons for which opinion I explain in the sequel. How then do we explain passages that ascribe the term 'substance' to such entities?

There are at least two explanations for why Aquinas does not always use the term 'substance' in the strict sense that I laid out in S4. The first explanation is that Aquinas thinks that 'substance' is a term that is used (rightly) in a number of different senses, senses which do not refer to entities that fall under a common genus. As Aquinas notes, x and y may be the same, although they are not numerically, specifically or generically the same. Rather, they can be the same by analogy.[40] 'Substance' is like many of the other key terms in Aquinas's metaphysical vocabulary in this respect. Although I cannot get into the complexities of Aquinas's theory of analogy here,[41] suffice it to say that Aquinas has such a theory, and this is one way of explaining why he sometimes employs the word 'substance' in a less than precise manner. Nevertheless, that Aquinas sometimes uses the term substance in this way does not mean that Aquinas does not think the term has a precise meaning or, at the very least, a primary meaning. Aquinas does think there is a primary sense, or most precise sense of 'substance'; this is the sense of substance that is given in S4.

A second explanation for loose talk where 'substance' is concerned, in particular with reference to the human soul, comes in seeing that many of Aquinas's theological authorities speak of the soul as a substance.[42] In fact, this explanation can be joined to the first one given above. Assume Aquinas's considered view about the human soul is that it is not a substance (in the strict sense). Aquinas need not be seen as disagreeing with Augustine when Augustine refers to the soul as a substance, since the soul is a substance, at least in an extended sense of 'substance', namely, insofar as it is a *hoc aliquid* or thing that can subsist on its own.[43]

Aquinas on the nature of material substances

Having said something about Aquinas's views on substance in general, I now want to say something about Aquinas's views on the nature of *material* substances in particular. Of course, Aquinas thinks that substances exist that are not made of any material; there are also immaterial substances in the world.[44] Before noting what is distinctive about material substances for Aquinas, it is worth exploring in greater detail what immaterial and material substances have in common.

For Aquinas, anything that actually exists is something having a certain form.[45] Although the meaning of the word 'having' in this statement needs to be qualified in the case of God (and perhaps for accidents as well), it holds true especially for created substances. In interpreting what Aquinas means by 'form', Eleonore Stump suggests employing the word 'configuration.'[46] In that case, a form is a configurational state of some sort.

Aquinas thinks there are two basic kinds of form, or configuration: substantial forms and accidental forms.[47] According to Aquinas, accidental forms are modifications of substances,[48] so that a substance's numerical identity does not change because it gains or loses such forms. As Aquinas puts it in several places, an accidental form makes an already existing substance to be something, not *simpliciter*, but in a certain respect (*secundum quid*).[49] For example, Socrates sits, and then later he stands up. He puts on a coat, but then decides to take it off. He gets sunburned, but that sunburn eventually fades. None of these changes make Socrates to be a different kind of thing, let alone cause him to go out of existence. Thus, such forms are accidental forms where Socrates's numerical identity is concerned. (The reader should note that 'accidental form' and 'accident' are interchangeable expressions for Aquinas. Therefore, whatever I have said about accidents above also applies to accidental forms.)

In contrast to accidental forms, substantial forms make substances to be in an absolute sense (*simpliciter*).[50] As Aquinas writes in one place, 'a substantial form gives being *simpliciter*, and therefore through its coming something is said to be generated *simpliciter*, and through its leaving corrupted *simpliciter*'.[51] Although Rex (the dog) can lose some of his forms without going out of existence, he cannot lose his substantial form without going out of existence. Thus, one characteristic of a substantial form is that it is a cause of a thing's actually existing.[52]

Substantial forms also explain why any given substance belongs to a particular species.[53] By 'species' in this context Aquinas means to refer to what medieval philosophers call a substance's 'lowest species' (*infima species*). The lowest species of a substance explains *what* that substance is, one might say, in the most fundamental way possible.[54] At any given time, Socrates has many different features, or configurations or forms. Some of these pick out *what* Socrates is, e.g. an animal, while others pick out *how* he is, e.g. white. However, to describe Socrates as an animal is not to describe what he is, most fundamentally (although it does partially describe what he is). Socrates is a human being, or a rational animal. 'Human being' is Socrates's *infima* species, since there is (arguably) no more fundamental way to express *what* Socrates is. It is Socrates's substantial form that confers on matter this sort of specific nature.

In the case of material substances, substantial forms configure something absolutely non-formal, namely, prime matter (matter that is not configured *per se*). In the case of immaterial substances – substances that do not have matter as a component part – their forms are themselves configurations.[55] The primary difference therefore between immaterial and material substances for Aquinas is that material substances have matter as a component part, while immaterial substances do not.[56]

For Aquinas, since something is material (or has matter as a component part) if and only if it is something extended in three dimensions,[57] a material substance would appear to be a substance that is three-dimensionally extended in space. I thus offer the following Thomistic definition of a material substance (MS):

> (MS) x is a material substance $=_{df} x$ (1) is a substance and (2) (normally) is three-dimensionally extended in space.[58]

But what should we say about (MS) given that Aquinas sometimes describes human beings as partly spiritual and partly corporeal (material)?[59] Since I take material substances and immaterial substances to be contradictory opposites, then it would be a problem for (MS) if human beings could not satisfy it (since human beings certainly are not immaterial substances for Aquinas). I will have more to say about some of the complications regarding the case of the nature of human beings in the sequel. For now let me say that human beings are indeed rightly classified as material substances for Aquinas, given the fact that Aquinas accepts 'rational animal' as a correct definition of 'human being'.[60] Furthermore, he thinks that the genus 'animal' is itself a species of the genus 'living substance', while 'living substance' is a species of the genus 'corporeal (or material) substance'.[61] Therefore, 'material substance' is implicit in the definition of the genus 'animal', and so also in the definition of the species 'human being'. Thus, whatever complications do arise for explaining the nature of human beings – in particular, explaining how human beings are a kind of animal that is capable of surviving biological death – it must remain true that human beings are correctly classified as material substances.[62]

Aquinas on the extension of 'material substance'

I have already shown that certain objects in our experience count as material substances for Aquinas, namely human beings. What other objects of our experience count as material substances? Aquinas offers what appears to be a comprehensive list of the different kinds of material substance in the context of discussing which of these will exist in the life to come. This list includes elements, compounds,[63] heavenly bodies,[64] plants, non-human animals and human beings.[65]

Among *corruptible* material substances, there are some that are non-living, while others are alive. It is clear what Aquinas has in mind in referring to plants and non-human animals. It is less clear to what Aquinas means to refer when he uses terms such as 'water' and 'bronze' to refer to material substances. (I will have more to say about Aquinas on human beings in Chapters 4 and 5.)

Broadly speaking, Aquinas distinguishes two sorts of non-living substance, elements and compounds. A compound is a type of non-living substance that is composed of a certain proportion of different kinds of elements. Aquinas clearly thinks that compounds are not mere aggregations of disparate elements, but substances in their own right.[66] Examples of compound-kinds for Aquinas include bronze,[67] silver,[68] copper,[69] stone,[70] blood,[71] flesh,[72] bone[73] and, apparently, cloud.[74]

The elements for Aquinas include earth, air, fire and water.[75] Aquinas offers a definition of 'element' in a variety of places. In his treatise *De principiis naturae*, Aquinas follows Aristotle in defining an element as 'that from which a thing is primarily composed, is in a thing, and is not divisible according to form'.[76]

Leaving to one side for now what Aquinas means by the elements being 'in' things,[77] let us first examine what Aquinas means by an element's being 'that from which a thing is primarily composed'. I take Aquinas to mean that element-kinds are substance-kinds whose instances are not in any way composed of bodies that differ from them in kind. To borrow expressions from Chapter 2, the element-kinds for Aquinas are ultimate mass-kinds or *homeomerous* mass-kinds. Recall my definition of a homeomerous mass-kind from Chapter 2:

(HO) K is a homeomerous mass-kind if and only if every mass x of K is such that, for any part p of x, p is some K.

Using HO as a guide, we can speak about the necessary and sufficient conditions for a homeomerous (elemental) *substance*-kind as follows:

(HS) A is a homeomerous (elemental) substance-kind if and only if for any x such that x belongs to A, if p is an integral part of x at time t such that p can be divided from x at $t + n$ and p is divided from x at $t + n$, then p at $t + n$ is either (i) identical to a substance y that belongs to A and y is not identical to x or (ii) an integral part $p1$ of a substance z that belongs to A and z is not identical to x and $p1$ is not identical to p.[78]

In explaining the meaning of Aristotle's claim that elements are not divisible according to form, Aquinas notes that elements are different from those things 'which have parts diverse in form, that is, in species, e.g. a hand, whose parts are flesh and bones, [things] which differ according to species'. Rather, for Aquinas 'an element is not divided into parts diverse according to species, e.g. water, each part of which is water'.[79] Aquinas appears to be making clearer here the way in which the elements are the primary constituents of things. Someone might think that Aquinas thinks of the elements as atomic in the sense that they are to be identified *only* with the smallest possible pieces of matter (the *minima*). But Aquinas does not think about the elements in this way. As he points out in another place, elements are not divisible into corporeal parts that differ in species, but they are nonetheless divisible into corporeal parts, that is, into like parts.[80] Any size portion of water counts as an instance of the elemental species 'water' for Aquinas.[81] And since elemental-kinds are substantial-kinds, any instance of an elemental kind (that is not currently a part of a substance) counts as a full-blown substance according to Aquinas. The elements are not to be identified only with the quantitative *minima*.

But if Aquinas thinks that the water in my bathtub and the piece of bronze that constitutes a statue count as instances of elemental and compound-kinds, respectively, does Aquinas believe that elemental and compound-kinds have *minima*, i.e. smallest portions of a given kind of matter?

Some texts give the impression that Aquinas does reject this tenet of atomism. For example, in the context of arguing that God is not corporeal, Aquinas says that every body is potentially something that it is not now, since bodies, as continuous, are divisible into infinity.[82] And in arguing that an actually infinite corporeal *magnitude* is impossible, Aquinas admits that a body may be *divided* into infinity.[83]

However, Aquinas notes in passages from his commentary on the *Physics* that there are limits to the degree that some material substances can be divided (by quantity) in actuality.[84] Some material substances cannot

actually be divided at all, even if they can be divided in thought, e.g. a heavenly body such as the sun cannot actually be divided (Aquinas supposes),[85]
but one can think about the side (or part) of the sun facing the earth and the
side (or part) of the sun that is not. Other material objects can actually be
divided, but certain of these quantitative divisions cause the object in question
to go out of existence, e.g. in the case of dividing in half organs such as the
heart.[86] Aquinas thinks that the integral parts of animals and plants have a
determinate range of sizes, because the wholes to which they belong have
such determinate ranges.[87] Furthermore, he notes that this is not at odds with
Aristotle's view that a continuum is divisible into infinity (an important worry
because elements and mixed bodies are continua for Aquinas). This is because
although there is a sense in which a body is divisible into infinity, that is, insofar as it is considered under the category of quantity, there is a sense in which it
need not be, namely, insofar as the body is considered as a natural substance
with a substantial form that requires that that substance have a determinate
species of quantity.[88] For example, Aquinas takes the nature of flesh to be such
that it has a determinate range of sizes, both in terms of how large and how
small a piece of flesh can be. Thus, although most visible pieces of flesh have
parts that are also pieces of flesh, any piece of flesh has some integral parts that
are not potentially pieces of flesh. If the smallest piece of flesh were quantita-
tively divided, then that piece of flesh would undergo an alteration that
resulted in its being resolved into its elemental parts. According to Aquinas,
compound-kinds clearly have minima (or '*K*-atoms,' to evoke terminology
from Chapter 2).

But it is less clear whether he thinks this is the case for all element-kinds.
It would appear that it is possible for the instances of some elemental kind
actually to be divided into infinity. For example, although Aquinas sometimes
comments in places that *water* has *minima*, his comments in those places need
not suggest that this is the case for all element-kinds.[89] Perhaps, for example,
fire is an element-kind whose every instance can be divided *ad infinitum*.[90]
Obviously, this is not something that could be accomplished by any feat of
human engineering, let alone by some natural process. But couldn't God
do it? If so, then Aquinas could reject the view that there are such things as
material atoms, whether such atoms be construed as partless entities or else
as indivisible entities that nonetheless have corporeal parts (which is, of
course, closer to the original Democritean sense of 'atom'). Nonetheless, even
if Aquinas does accept that there are *minima* for every non-living substance
kind – including homeomerous substance kinds – it nonetheless should be
remembered that Aquinas does not take the *minima* of a substance kind to
be the only instances of that kind. A visible portion of water is every bit as
much a substance and an element for Aquinas as those instances of water that
are *minima*.

There is a final question that I want to address at this point concerning Aquinas's views about the extension of 'non-living, material substance': which compound-kinds have *substances* and not merely *parts of substances* as instances? Some compound-kinds clearly have instances that are substances for Aquinas, e.g. *bronze*.[91] But do all compound-kinds have substances, in the strict sense, as instances according to Aquinas? It would seem that there is good reason for thinking that some compound-kinds do not have substances as their instances. Consider compound-kinds such as blood, flesh and bone. Unlike compound-kinds such as bronze, these appear to get their species from the material substances of which those compounds are (normally) integral parts. In other words, these kinds are such that every one of their instances is a *hoc aliquid* that is incomplete in its species. Take as an example the flesh that currently exists as part of your hand. Now imagine an unfortunate accident where your right hand is severed from your body. After a certain period of time (or perhaps as soon as it is severed from your body) that thing that you refer to as 'your right hand' is a hand in an equivocal sense only according to Aquinas.[92] For reasons to be discussed in the sequel, what one continues to refer to as a 'hand' after it has been severed from a human body is just as much a real hand as those portions of Michelangelo's *David* that are shaped like human hands. But one might think that if your hand gets its species from being a part of a living human body, then so does the flesh that is a part of your hand. In that case, the flesh that is severed from your body is itself no longer flesh (although it may look a lot like flesh). That would mean that compound-kinds such as flesh do not have substances as their instances, and that not all compounds are substances for Aquinas; some kinds of compounds always exist as parts of substances only and do not really have substances as instances.

Noticeably absent from Aquinas's list of substances are artefacts. Why does Aquinas not consider artefacts to be substances? In order to answer this question, I need to say more about Aquinas's understanding of the different ways that material objects are composed of parts. The next chapter therefore addresses Aquinas's views on composition and part-hood.

Notes

1. See, for example: Lowe 1998 and Hoffman, J. and Rosenkrantz, G. (1997) *Substance: Its Nature and Existence*, London: Routledge.
2. See, e.g., In Met. VII, lec. 1. For Aquinas's reasons for thinking that being cannot be a genus, see, e.g., ST Ia. q. 3, a. 5.
3. See, e.g., QDP q. 7, a. 3 ad4.
4. See, e.g., QDP q. 7, a. 3: 'Sed si substantia posit habere definitionem, non obstante quod est genus generalissimum, erit eius definitio: quod substantia est

res cuius quidditati debetur esses non in aliquo.' See also, In Sent. IV, d. 12, q. 1, a. 1, ql. 1, ad2; ST Ia. q. 3, a. 5, ad1 and ST IIIa. q. 77, a. 1, ad2.

5. Aquinas discusses four meanings of 'substance' at In Met. V, lec. 10, nn. 902–4 and In Met. VII, lec. 2, nn. 1270–4. However, he thinks these four senses of 'substance' can be reduced to two, namely substance as essence, or quiddity, and substance as primary substance. See, for example, In Met. V, lec. 10, n. 904; ST Ia. q. 29, aa. 1 and 2; ST IIIa. q. 2, a. 6, ad3; ST IIIa. q. 17, a. 1, ad 7 and SCG IV ch. 49. For a comprehensive look at Aquinas's discussions of the different senses of 'substance' and his reduction of these four senses to two, see: Wippel, J. (2000) *The Metaphysical Thought of Thomas Aquinas: From Finite Being to Uncreated Being*, Washington, DC: The Catholic University of America Press, pp. 201–8.

6. See, e.g., In Met. V, lec. 9, n. 891.

7. See, e.g., QQ 9 q. 2, a. 1, c. and ST Ia. q. 29, a. 1, c.

8. See, e.g., In Phys. I, lec. 11, n. 92 and In Met. IV, lec. 1, n. 539.

9. See, e.g., Descartes's *Meditations on First Philosophy*, Third Meditation, sec. 44: 'a substance . . . is a thing capable of existing independently' (1996, p. 30). Defining substance as non-dependent being still holds sway for many contemporary philosophers. See, e.g., Lowe 1998, p. 151 and Armstrong, D. (1978) *Nominalism and Realism*, Cambridge: Cambridge University Press, p. 115.

10. See, e.g., SCG I chs. 13–28 and ST Ia. qq. 2–4. Hoffman and Rosencrantz are two contemporary philosophers who build their own definition of substance mindful of the fact that some philosophers take God to be the cause of contingent substances (1997).

11. See, e.g.: SCG I ch. 25 (10); QDP q. 7, a. 3 ad4 and ST Ia. q. 3, a. 5, ad1. Cf. In Sent. I, d. 8, q. 4, a. 3, ad1. Here Aquinas says that 'substance' can be predicated of God and creatures, albeit analogously and not univocally.

12. See, e.g.: DPN ch. 1; In Met. V, lec. 10, n. 903; and In Met. VII, lec. 2, nn. 1291–3.

13. I owe this insight to Bobik, J. (1965) *Aquinas on Being and Essence: A Translation and Interpretation*, Notre Dame, IN: University of Notre Dame Press, p. 41.

14. See, e.g.: In Sent. I, d. 8, q. 4, a. 2, ad2; In Sent. II, d. 3, q. 1, a. 5, c.; In Sent. IV, d. 12, q. 1, a. 1, ql. 1, ad2; QQ 9 q. 3, a. 1 ad2; SCG I ch. 25 (10); QDP q. 7, a. 3 ad4; ST Ia. q. 3, a. 5 ad1 and ST IIIa. q. 77, a. 1, ad 2.

15. See, e.g., ST Ia. q. 3, a. 4.

16. See, e.g., QQ 9, q. 3, a. 1, ad2 and ST IIIa. q. 77, a. 1, ad2. Note that I am construing accidents as individuals, which is how Aquinas conceives of them (see the explanation for this view of Aquinas's in the sequel).

17. See, e.g.: Plantinga, A. (1974) *The Nature of Necessity*, Oxford: Clarendon Press, pp. 60–9.

18. See, e.g.: In Sent. I, d. 3, q. 4, a. 2, c.; ST Ia. q. 76, a. 6, ad1; ST Ia. q. 77, a. 1, ad5 and ST Ia. q. 77, a. 6, c. Elsewhere, Aquinas refers to proper accidents as '*necessaria*' (DPN, ch. 2), '*inseparabilia accidenta*' (In Phys. I, lec. 6, n. 45), and '*per se accidenta*' (ST Ia. q. 77, a. 6, c.).

19. See, e.g., ST Ia. q. 77, a. 6, c. Elsewhere, Aquinas refers to extraneous accidents as '*non necessaria*', '*communia accidenta*' (In Sent. I, d. 3, q. 4, a. 2, c.) and '*separabilia accidenta*' (DPN ch. 2).

20. See, e.g.: DPN ch. 2 (7); In Met. V, lec. 22, nn. 1142–3; ST Ia. q. 76, a. 6, ad1 and ST Ia. q. 77, a. 1, ad5. I am indebted to Eleonore Stump and Alicia Finch for help with the wording of this definition.

    I am not sure whether Aquinas thinks that only substances belonging to kind *A* have proper accidents belonging to *F*. If this is the case, then the following clause needs to be added to the *definiens* of PA: (4) for any *a* such that *a* does not belong to *A*, there is no accident belonging to *F* that inheres in *a*.

21. See, e.g., ST Ia. q. 44, a. 1, ad1. My definition of a proper accident entails that a human being could not lose the ability to laugh without going out of existence for Aquinas. However, I am not sure this is right. It could be the case that, for Aquinas, a thing could lose a proper accident without going out of existence. For example, proper accidents might be instances of those kinds of accidents that an instance of a kind of substance *normally* has. One metaphysical issue relevant to resolving this interpretive issue is Aquinas's views on the relation between composition and identity. See Chapter 6 for discussion.

22. See, e.g.: Yablo, S. (1998) 'Essentialism' in Edward Craig (ed.) *Routledge Encyclopedia of Philosophy*, vol. 3, London: Routledge, p. 417, and Van Cleve, J. (1995) 'Essence/accident' in J. Kim and E. Sosa (eds) *A Companion to Metaphysics*, Oxford: Blackwell, pp. 136–8.

23. See, e.g., ST Ia. q. 77, a. 1, ad1.

24. See, e.g., DPN ch. 2 (7) and In Met. V, lec. 22, nn. 1139–41.

25. See, e.g.: Reith, H. (1958) *The Metaphysics of St Thomas Aquinas*, Milwaukee, WI: Bruce Publishing, p. 87.

26. The term 'predicamental' comes from Aquinas's '*praedicamentum*', which itself is the Latin translation of Aristotle's κατηγορία. For some places where Aquinas speaks about the *praedicamenta*, see, e.g.: SCG II ch. 9; ST Ia. q. 5, a. 6, ad1; and ST Ia. q. 28 a. 2, ad1.

27. See, e.g.: DPN ch. 1; In Sent. I, d. 8, q. 5, a. 2, c.; ST Ia. q. 9, a. 2, c.; In Phys. I, lec. 2, n. 14; In Phys. I, lec. 12, n. 108; In Phys. II, lec. 1, n. 145; In Met. V, lec. 9, nn. 885–7; In Met. VI, lec. 2, n. 1179 and In Met. VI, lec., 4, nn. 1241–4. See also SCG II ch. 57 (3), where Aquinas uses the expression, *ens per accidens*. See also DEE ch. 7, SCG II ch. 58 (6), ST Ia. q. 76, a. 3, c., and In Met. V, lec. 7, nn. 842ff., where Aquinas speaks about the composition of a substance and a non-necessary accident as something *unum per accidens*, and texts such as ST Ia. q. 5, a. 1, ad1 where Aquinas speaks of a substance as having *ens simpliciter* and the composition of a substance and an accident (a kind of accidental being) as having *ens secundum quid*.

    Aquinas also uses 'accident' in at least two other senses that I do not speak about here (accident as a predicable, and accident as something incidental to a subject of inquiry or intention).

28. I will have more to say about these different kinds of parts in Chapter 4.

29. A material form is a kind of *substantial* form – a form whose presence causes a substance to be and to be the kind of thing that it is most fundamentally. Specifically, a *material* substantial form is a substantial form that does not rise above matter in its nobility so as to have an operation that functions independently of matter. See the sequel for more discussion of material and substantial forms.

30. See, e.g., ST Ia. q. 29, a. 2, c. and QQ 9 q. 2, a. 1, c. For Aquinas, whatever exists extra-mentally is particular. See, e.g., DEE ch. 4.

31. ST Ia. q. 75, a. 6.

32. See, e.g., In Met. VII, lec. 1, nn. 1248–51. As Joseph Bobik notes, we do not hesitate to refer to some human being named 'Jack' as a 'being'. However, this is not the case when it comes to Jack's height at some time. Although Jack's height '*is there* . . . as opposed to *not there at all*', Jack's height is not there 'in the sense in which we say that Jack is there, i.e. simply or without qualification. We would want to make a qualification; we would say it is there *in* something which is there simply, in this case *in* Jack. The word "in" would mean *as a modification* or *characteristic of*' (1965, pp. 32–3).

33. See also: ST Ia. q. 75, a. 2, ad2; ST Ia. q. 75, a. 4, ad2 and QDA a. 1, ad3.

34. See, e.g., DEE ch. 7.

35. See also In DA II, lec. 1, n. 213.

36. See, e.g., ST Ia. q. 3, a. 5.

37. See, e.g.: SCG II ch. 68; ST Ia. q. 75, prologue and ST Ia. q. 75, a. 4, ad2.

38. See, e.g., ST Ia. q. 75, a. 4, ad2.

39. See, e.g., In DA II, lec. 1, n. 218.

40. See, e.g., DPN ch. 6 (32) and (35).

41. For a good presentation of Aquinas's theory of analogy, see: McInerny, R. (1996) *Aquinas and Analogy*, Washington DC: Catholic University of America Press.

42. See, e.g., ST Ia. q. 75, a. 2, *sed contra* (where he cites Augustine) and ST Ia. q. 75, a. 6, *sed contra* (where he cites Pseudo-Dionysius).

43. See, e.g., Aquinas's gloss on Augustine at ST Ia. q. 75, a. 2, *sed contra*.

44. Aquinas offers arguments for the existence of immaterial substances at ST Ia. q. 51, a. 1, c and QDSC a. 5, c. Of course, Aquinas also takes the witness of the Old and New Testaments seriously when it comes to the reality of immaterial substances such as angels.

45. DPN ch. 1 (3).

46. Stump, E. (2003) *Aquinas* (Arguments of the Philosophers series), London: Routledge, p. 36.

47. See, e.g., DPN ch. 1.

48. See, e.g.: DPN ch. 1; DEE ch. 7, and; ST IIIa. q. 77, a. 1.

49. See, e.g.: DPN ch. 1 (4); DEE ch. 2 (18), and; ST Ia. q. 76, a. 4.

50. See, e.g.: SCG II ch. 68 (3); ST Ia. q. 50, a. 5, c. and ST Ia. q. 77, a. 1, ad3.

51. ST Ia. q. 76, a. 4, c.: 'Forma autem substantialis dat esse simpliciter: et ideo per eius adventum dicitur aliquid simpliciter generari, et per eius recessum simpliciter corrumpi.'

    Although I have certainly employed existing English translations of Aquinas's works (noted in the Select Bibliography) as guides throughout, all translations of Aquinas's works are my own.

52. See, e.g., DPN ch. 1 (3) and DME (11). This claim should not be taken to exclude the possibility that Rex depends upon other factors in order to exist in the first place, or to remain in existence once he exists *simpliciter*. Indeed, Rex must have some matter as a component part and two parents in order to exist, let alone a first

and sustaining cause (God) according to Aquinas. Nevertheless, the presence of a substantial form is a sufficient condition for a substance being in existence. The point might be made in another way. Whereas material substances have a number of different causes, e.g. material, final, productive and formal, the substantial form is a material substance's intrinsic formal cause (see, e.g., SCG II ch. 68 [3]).

53. See, e.g.: DPN c. 1; In DA II, lec. 1, nn. 213, 215; CT ch. 21; QDA 1; QDA 9; QDSC a. 1, ad 9; QDSC a. 3; ST Ia. q. 76, a. 2; ST Ia. q. 76, a. 4 and In Met. VIII, lec. 1, nn. 1688–9.

54. Cf. Lynne Rudder Baker's view that every material object belongs to a *primary kind* (2000, pp. 39–41).

55. As Stump explains this position, 'an angel has no matter to configure, but it is nonetheless configured in a certain way. It has certain properties, such as being a knower, and not others, such as weighing two hundred pounds. And so there is a kind of configuration in an angel, too, which we can think of as an organization of properties.' (2000, p. 37).

56. See, e.g.: SCG II ch. 50; SCG IV ch. 84 (11); ST Ia. q. 50, a. 2, c. and QDSC a.1, c.

57. ST Ia. q. 3, a. 1, ob. 1; ST Ia. q. 52, a. 1, c., and ST Ia. q. 52, a. 2, c.

58. Including the term 'normally' here allows MS to be consistent with the complication in Aquinas's views that human beings sometimes exist in a state such that they do not have matter as a part.

59. See, e.g., ST Ia. q. 50, prologue and ST Ia. q. 75, prologue.

60. See, e.g.: ST Ia. q. 2, a. 1, c.; ST Ia. q. 13, a. 12, c.; ST Ia. q. 76, a. 3, c. and ST Ia. q. 76, a. 3, ob. 4. See also SCG II chs. 57 (5) and 58 (3 and 7), where Aquinas says that humans are animals essentially.

61. See, e.g., SCG II ch. 65 (2).

62. See also SCG II ch. 57 (5) and SCG II ch. 58 (3).

63. I use the English locutions 'mixed bodies' and 'compounds' interchangeably to translate Aquinas's expression, '*mixta corpora*'.

64. The heavenly bodies differ from other kinds of material substances in that they enjoy the unique characteristic of being incorruptible (see, for example, ST Ia. q. 50, a. 5, c.; ST Ia. q. 50, a. 5, ad 3; ST Ia. q. 66, a. 2, (end of) c.; ST Ia. q. 68, a. 4, c., and; ST Ia. q. 70, a. 3, c.).

65. CT ch. 170. See also, SCG II ch. 68 (7–12) and ST Ia. q. 76, a. 1, c.

66. In addition to those texts cited in note 65 above, see also DME and CT ch. 211.

67. See, e.g., DPN 7.

68. See, e.g., In Met. V, lec. 3, n. 779.

69. See, e.g., In Met. V, lec. 4, n. 796.

70. See, e.g.: ST IaIIae. q. 6, a. 1, c.; CT ch. 211; and In Met. V, lec. 10, n. 898. In the passage from the ST, Aquinas speaks of a stone (*lapis*) as having a principle of motion intrinsic to it. This indicates that he thinks that the stone itself is a substance. See In Phys. II, lec. 1, n. 146.

71. See, e.g., In Met. V, lec. 10, n. 898.

72. See, e.g., In Met. V, lec. 10, n. 898.

73. See, e.g., DPN 3 (19).

74. See, e.g., CT ch. 154.

75. See, e.g., CT ch. 170. For an interesting discussion of each of the Thomistic elements, see Bobik, J. (1998) *Aquinas on Matter and Form and the Elements*, Notre Dame, IN: University of Notre Dame Press, pp. 167–83.

76. DPN 19: 'Unde Aristotles in V *Metaphysicae* dicit quod elementum est id ex quo componitur res primo, et est in ea, et non dividitur secundum formam.' See also In Met., V, lec. 4, nn. 795–8 and 800.

77. I discuss what Aquinas means by elements being such that they exist *in* compounds whenever they come to compose compounds in Chapter 4.

78. Note that HS is compatible with both the truth and falsity of the view I labelled 'atomism' in Chapter 2 – the view that says there are instances of *A* that do not have proper parts (what we might call 'a-atoms').

79. DPN 19: 'Tertia particula, scilicet *et non dividitur secundum formam*, ponitur ad differentiam eorum quae habent partes diversas in forma, idest in specie; sicut manus cuius partes sunt caro et ossa, quae differunt secundum speciem. Sed elementum non dividitur in partes diversas secundum speciem, sicut aqua, cuius quaelibet pars est aqua.'

80. In Met. V, lec. 4, n. 800. See also In V Meta., lec. 4, n. 798 where he makes it clear what he means by 'not divisible into different species' by contrasting elements with compounds, e.g. blood, any instance of which can be divided into components that differ from blood in species, i.e. into the elements that compose any portion of blood.

81. See, e.g.: ST Ia. q. 3, a. 7, c.; ST q. 11, a. 2, ad2 and ST Ia. q. 119, a. 1, ad 5.

82. ST Ia. q. 3, a. 1, c.

83. ST Ia. q. 7, a. 3, c. See also In Phys. III lec. 1, n. 277 and ST Ia. q. 7, a. 1, ad2.

84. In Phys. VIII, lec. 11, n. 1067.

85. I presume that Aquinas means here that a heavenly body cannot be actually divided by *natural* causes. Certainly God could divide a heavenly body if he so chose.

86. See also In DA II, lec. 8, n. 332, where Aquinas argues that by nature things have limits with respect to how large and small they can be. However, he seems to be speaking primarily of living substances in this passage.

87. In Phys. I, lec. 9, n. 65.

88. In Phys. I, lec. 9, n. 66. See also QDP q. 4, a. 1, ad5 and ST Ia. q. 68, a. 2, (the very end of the) c.

89. QDP q. 4, a. 1, ad5 and ST Ia. q. 68, a. 2, (the very end of the) c.

90. But see In Phys. I, lec. 9, n. 65, In Phys. I, lec. 9, n. 66, QDP q. 4, a. 1, ad5, and ST Ia. q. 68, a. 2, (the very end of the) c.

91. See, e.g., DPN ch. 1

92. See, e.g., ST Ia. q. 76, a. 8. See Chapter 4 for further discussion.

4

Aquinas on the Composition of Material Objects

In the last chapter I explained what Aquinas has in mind by those material objects he refers to as 'substances'. In this chapter I want to focus on Aquinas's account of the parts of material objects, and the parts of material substances in particular.

This chapter has three sections. First, I explain Aquinas's views on the different ways that material objects are composed. Second, I offer a detailed analysis of those modes of composition that are especially relevant for offering Thomistic solutions to the classical puzzles having to do with the composition of material substances. I say *modes* (in the plural) of composition, since Aquinas thinks that the nature of material substances requires one to say that they are composed in a number of different ways. Having explained Aquinas's views on the composition of material substances, I close the chapter with a section that addresses Aquinas's views on the nature of artefacts and artefact composition.

Aquinas on the different ways that material objects are composed[1]

Aquinas speaks about composition in a variety of different ways, and in a variety of different contexts throughout his corpus. For example, when he is treating the simplicity of God, Aquinas argues that God is absolutely simple since there is no sense in which God is a *compositio*.[2] Also, in commenting on Aristotle's *Metaphysics* Aquinas offers detailed treatments of the different ways that Aristotle speaks about composition.[3] Although I will have occasion to treat other texts from his corpus in the sequel that are related to our topic, I turn my focus now on a few places where Aquinas treats the different modes of composition in his two great systematic works – *Summa theologiae* and *Summa contra gentiles*. Texts in these two works offer us a comprehensive look at Aquinas's views on the different ways that material objects in particular are composed.

For example, Aquinas distinguishes two ways that many things can be united (*modi possunt uniri*) in the context of treating the question how an intellectual substance could be united to a body.[4] Aquinas says that things can be united by way of mixture (*per modum mixtionis*) – so as to form a compound – or things can be united by way of contact – whether in the sense that two bodies are in contact at their extremities, or in the sense of contact by power, as, for example, when an immaterial substance influences the position of a body without physically touching it.

In speaking about mixtures in this passage from SCG II, Aquinas also mentions a kind of composed object that he refers to as a *confusio*, which might be translated 'collection' or 'aggregate'. In contrast to a compound (*mixtio*), whose integral parts are present by power, or potentially and not actually, the parts of a *confusio* are present actually. In other words, the parts of a compound are not actual or complete substances while they are parts of that compound, whereas the parts of a *confusio* are actual or complete substances. As we shall see, for Aquinas this entails that a *confusio*'s form of the whole is an accidental and not a substantial form – that the *confusio* is not itself an actual substance.

Aquinas also talks about composition by listing the different ways that many things are made one. So, for example, in the context of treating the union of Christ's two natures in SCG IV, Aquinas suggests that there are a number of different ways that something one in nature or species is made from many things.[5] He says that many things are made one by order alone (*secundum ordinem tantum*), e.g. as a city is made up of many homes, by order *and* composition, e.g. as walls and roof come together to form a house, and through a co-mingling (*per commixtionem*), e.g. as a compound is made up of its constitutive elements. He implies later in the passage that composition by form and matter represent another way many things are made one, and, finally, the way an animal is composed of its integral parts counts as still another kind of composition.

A similar (but slightly different) list of ways that many things are made one can be found in a passage from the ST, a passage which too is treating questions about the mode of the union of Christ's two natures.[6] Here Aquinas talks about one thing being constituted by two or more things under three broad categories. The first category includes forms of composition where the parts of the composite in question remain *perfect*, that is, they exist as actual substances all the while they remain parts of the composite. Aquinas posits two species of composition within this category: *composition with order and figure* (or by way of assemblage [*per modum commassationis*]), e.g. as a house's parts compose a house, and *composition by way of aggregation* (*per modum confusionis*) or without order, e.g. as a heap of stones is composed of many stones. Both of these species of composite have instances that are accidental beings, and not substances, since their forms of the whole are accidental and not substantial forms. As Aquinas also puts it, such composites are not one *simpliciter*, but rather one *secundum quid*.

Aquinas's second way – categorically speaking – that many things are made one picks out things composed of parts that are perfect and yet transformed. Here Aquinas has in mind the way elements continue to exist in a compound composed of those elements. As we shall see in the sequel, elemental parts differ from both integral parts and metaphysical parts for Aquinas.

Finally, a third categorical way that many things are made one includes those things that are composed of imperfect entities, e.g. as a substance (a perfect entity) is composed of form and matter (imperfect entities). Here we have Aquinas mentioning composition out of metaphysical parts.

The passages I have examined above from the SCG and ST provide us with a fairly comprehensive list of the different ways Aquinas thinks that material objects are composed. Combining the different kinds of composition listed in these parallel passages from the SCG and ST, we can list the following *four broad categories of composition* where material objects are concerned: (1) composition out of parts that remain actual substances in the composite. For convenience of expression, I will refer to this category of composition as 'accidental composition'; (2) composition out of integral (or quantitative) parts, e.g., the way an animal is composed of a head, torso, feet, etc.; (3) composition out of parts that remain (in some sense) perfect, while yet being transformed, i.e. compounds insofar as they are composed of the elements; and (4) composition out of imperfect entities, i.e. a material substance's composition out of substantial form and prime matter.

Furthermore, the category of accidental composition itself has three species of composition falling under it: 1a) things made one by order alone, e.g. an army, 1b) things made one by order and composition, e.g. a house built of walls and a roof, and 1c) many perfect things made one through aggregation, e.g. a heap of stones.[7]

A question arises as to the logical relationship between the four broad categories of composition where material objects are concerned. It appears that these categories are not species of one genus, e.g. the genus 'composition', nor are any of these categories themselves a genus for the others. Integral parts are three-dimensional entities, in contrast to metaphysical and elemental parts. And, although the integral parts of accidental composites are three-dimensional, in contrast to the integral parts of substances – which are not substances for Aquinas – the integral parts of accidental composites are substances. Also, although elemental parts are clearly material components of a compound, they should not be confused with matter taken as a metaphysical part. These differences between the different kinds of parts in material objects will become clearer when I examine them one by one in the next section. At this point I want to suggest the possibility that these four different kinds of composition are akin to Aristotle's ten categories: these different kinds of composition do not have a common genus, even though we use the same words to describe them, i.e. 'part', 'whole', 'composition' and 'composite'. It appears as if these terms are related by way of analogy, and not by species to one overarching genus.[8]

Aquinas on the composition of material substances

My focus in this section is on the different ways that material *substances* are composed. I begin the section with Aquinas's account of what one might refer to as the 'metaphysical composition' of a material substance, or the way in which a material substance is composed of metaphysical parts. I then go on to discuss Aquinas's understanding of the quantitative parts of material substances that Aquinas typically refers to as 'integral parts'. I conclude the section with a brief discussion of Aquinas's views on how elements remain in compounds as parts.

Metaphysical parts: form/matter composition in material substances

Aquinas considers the form and matter of a material substance to be parts (of a certain sort) of that substance.[9] Just what these 'metaphysical'[10] parts are and therefore what Aquinas has in mind by composition out of metaphysical parts will become clearer once one understands what Aquinas means by 'substantial form' and 'matter'.

Aquinas on substantial form

As we saw in Chapter 3, Aquinas thinks that a material substance's substantial form causes that substance to be *simpliciter*. Of course, that a substantial form causes a substance to be *simpliciter* does not entail that a material substance has no extrinsic causes, such as God, but only that, if a substantial form of a certain species *A* is present in matter, then that is sufficient for there being an actually existing material substance of the species *A*. In addition to this role of causing a material substance to be *simpliciter* is the substantial form's role of causing a substance to belong to its (*infima*) species. As Aquinas writes, 'one and the same form, through its essence, makes a human being an actual being, a body, a living thing, an animal, and a human being'.[11] Furthermore, the substantial form of a substance causes that substance to have its proper accidents, such as risibility in human beings. The only features of a substance that are not explained by the presence of a substance's substantial form are its extraneous accidents, e.g. the whiteness of Socrates's skin.[12] Although Socrates's having the substantial form of a human being is a necessary condition for having fair skin – since his substantial form confers on Socrates the property of having a body that is capable of being coloured in the first place – it is not a sufficient condition: Socrates's skin being white also depends upon factors extrinsic to Socrates's substantial form such as the climate in which Socrates lives and Socrates's particular human lineage.

The causal powers of a substantial form also extend to the integral parts of a substance for Aquinas. As Aquinas writes in one place:

> The substantial form is not the perfection of the whole alone, but of each part. For since the whole consists of parts, a form of the whole that does not give being to the particular parts of a body is a form that is [mere] composition and order, as in the form of a house. [But] a form of this kind is accidental. However, a soul is a substantial form; hence it must be that it is the form and act not of the whole alone, but of each part [of the whole].[13]

As the reader will note, it is not because a form is a soul that it causes the parts of a substance actually to exist, but because that form is a substantial form. What Aquinas says here about souls holds true for all substantial forms. Thus, we might say that insofar as a substance has any essential features, this is due to the presence of that substance's substantial form.

So far I have described what Aquinas takes a substantial form to be by describing its effects. But just what *is* the substantial form of a material substance? With the exception of one sort of substantial form (the human soul), the substantial form of a material substance is not itself a *being* but that by which (*a quo*) something is a being.[14] As we'll see, the same holds true for that other metaphysical component of a material substance, prime matter.[15]

Another way of making the point that a substantial form is not a being is to say, as Aquinas sometimes does, that substantial form is an intrinsic principle or cause[16] of being (but not itself a being).[17] For Aquinas, if *x* is a principle or cause of *y*, *x* is an ultimate explanation of some feature of *y*.[18] Now Aquinas thinks that every material substance has two *intrinsic* principles (or causes): a substantial form and prime matter. As Aquinas points out, just because *x* is a principle of *y*, this does not entail that *x* is a part of *y*. Thus, my parents are, in one sense, a productive cause of me, and therefore they function as principles of my being. But my parents are not parts of me (at least not in a literal sense of 'part'). My substantial form is also one of my principles according to Aquinas, but in contrast to my parents, my substantial form *is* a part of me (in a literal sense of 'part').

A substantial form is a *metaphysical* part: it explains why that substance actually exists and is the sort of substance that it is. And just like more garden-variety parts of material substances – parts that Aquinas calls 'integral parts', e.g. a hand – metaphysical parts of material substances are not themselves substances or accidents. It makes sense to call them *parts* of a substance.

That prime matter and substantial forms are not substances is one of the reasons Aquinas says that the union of a substantial form with matter produces *one* being, and not simply the correlation of many beings. Thus, in a composed object that has pre-existing things as parts, Aquinas argues that it is necessary that something bind such parts together.[19] This is not so for composites of form and matter since form/matter composition does not involve the putting together of actual, pre-existing things. As Aquinas writes,

> Form is united to matter without any medium; for it belongs to a form in virtue of itself that it is the act of a kind of body, and not through something else. Hence, nothing [besides] matter and form makes something one except the agent that reduces the potency [of matter] to act, as Aristotle proves in the eighth book of the *Metaphysics*. For matter and form are related as what is potential to what is actual.[20]

Aquinas makes two points here. It is in the very nature of the substantial form of a material substance to be united to matter (and, as we shall see, matter does not actually exist unless it is configured by a substantial form). Thus, and this is the second point, substantial forms do not exist prior to configuring matter. Substantial forms and prime matter are not parts that actually pre-exist the composite to which they belong. Because matter and form are principles of being and not themselves beings, they compose something that is one *simpliciter*, and do so without something intrinsic to the composed thing binding them together.

Although substantial forms are not beings, properly speaking, but rather principles of beings, one would be wrong to conclude that the distinction between substantial form and matter in a material substance is merely a logical one. That is to say, for Aquinas the distinction between a substantial form and prime matter in a material thing is not merely a distinction of reason. Substantial forms and prime matter are real constituents of a material substance, although they are not real in the same sense that substances (or accidents) are.[21]

As we saw above, Aquinas makes the claim that 'matter and [substantial] form are related as what is potential to what is actual'. What does it mean to say that form is something actual? As we have seen, this cannot mean that substantial forms are beings whereas matter is not a being. Like matter, some (but not all) substantial forms do not exist as beings. It is the composite of substantial form and matter that is properly speaking a material being, that is, a material substance.

It will be helpful at this point to return to the notion that forms are configurational states.[22] Unlike accidental forms, substantial forms configure prime matter and not some already existing substance. Form is something actual in relation to matter since, as Joseph Bobik puts it, 'a composed substance does not begin to be until the form is introduced'.[23] In other words, matter without the substantial form of something belonging to species *A* does not exist as a part of something belonging to species *A*. The matter in question becomes a component of something belonging to species *A* at the moment it receives a substantial form of the appropriate sort. So although such a substantial form does not actually exist in extra-mental reality before it configures matter, the substantial form, more than the matter that receives it, explains why a substance actually exists as a member of species *A*. The matter in such a substance can be the matter of any variety of material substances; this is not so with the substantial form. A substantial form configures matter so that it is this kind of substance rather than another kind. This is why Aquinas says that the substantial form gives actual being to this substance (being *simpliciter*), and why he says that a substance's species is derived from its substantial form.[24]

Aquinas's views on substantial form thus show that he is not thinking of substantial forms as equivalent to the spatial relationships that exist among that substance's material parts. This is because the substantial form of a material substance causes the parts of a substance to be what they are, and thus to be in the places that they are. Since something cannot be the cause of itself, a substantial form cannot be equivalent to the spatial relationships that obtain among a material substance's integral parts. Thus, the substantial form of a substance exists over and above a substance's integral parts, giving such parts more than a mere aggregative unity.[25] As Aquinas goes on to say with respect to the composite whole that is one thing *simpliciter*, e.g. a substance: 'The composite itself is not [identical] to those things from which it is composed.'[26] To describe Aquinas's view here in a contemporary idiom, one might say that, for Aquinas, 'constitution is not identity'.

Let me make one last observation about Aquinas's general views on substantial forms. The substantial forms of material substances are individual things for Aquinas, and not universals, as some have wrongly supposed.[27] Not only does Aquinas make this point in many places,[28] but that Aquinas takes substantial forms to be individuals and not universals can be seen from the fact that the substantial forms in which he takes the most interest, namely human souls, are construed by Aquinas as individual things. Nor is this latter point softened by the fact that human souls are unusual sorts of substantial forms in that they can exist apart from matter. As we shall see in the sequel, Aquinas's view that human souls are subsistent entities does not conflict with the fact that they really are substantial forms. Thus, there is no reason to think that the individuality of the human soul is an exception to some general rule. Indeed, that substantial forms are individual things and not universals comports with Aquinas's general view with respect to universals, namely, that they exist only in the mind. For Aquinas, anything that exists extra-mentally is an individual.[29]

Having said these things about substantial forms, let me propose the following definitions, first of 'metaphysical part' and then of 'substantial form (of a material substance)':

(MP) *x* is a metaphysical part of a substance $=_{df}$ (1) *x* is a principle of *y*'s being *simpliciter*, and 2. *x* is something intrinsic, rather than extrinsic, to *y*.

(SF) *x* is a substantial form (of a material substance) $=_{df}$ (1) *x* is a metaphysical part, and (2) for any *y* such that *x* is a metaphysical part of *y*, *x*'s being a part of *y* is sufficient to make it the case that *y* exists as an instance of some species *A*.

Different kinds of substantial form

According to Aquinas, there are different kinds of substantial forms, since there are different kinds of material substances. One way that Aquinas some-times speaks about the differences between substantial forms is in terms of their relative perfection. For Aquinas a kind of substantial form is the more perfect to the extent that the features, powers and operations it confers on a substance are – to use a contemporary idiom – 'emergent', that is, are fea-tures of a substance that cannot be said to belong to any of the integral parts of the substance that is configured by that substantial form.

> It must be considered that the more noble a form is, the more it rises above [*dominatur*] corporeal matter, and the less it is merged in matter, and the more it exceeds matter by its operation or power. Hence, we see that the form of a mixed body has a certain operation that is not caused by [its] elemental qualities.[30]

A substance's substantial form is something above and beyond the properties of that substance's integral parts. What's the proof? Substances have powers and operations that are not identical to any of the powers and operations of that substance's integral parts, taken individually. Nor are the powers con-ferred by a substantial form of a substance *x* identical to a mere summation of the powers of the integral parts of *x*. So, a mixed body such as bronze has cer-tain powers that none of its elemental parts have by themselves, nor when those parts are considered as a mere sum.[31]

Consider that substantial forms fall into the following sort of hierarchy of perfection. The least perfect kind of substantial form corresponds with the least perfect kind of material substance, namely an element. Aquinas says that the substantial forms of the elements are wholly immersed in matter (*tota-liter immersae materiae*), since the only features that elements have are those that are most basic to matter (*qualitates quae sunt dispositiones materiae*).[32] In contrast, the substantial forms of compounds have operations that are not caused by their elemental parts. Above the forms of compounds, the forms of plants reach a level of perfection such that they get a new name: 'soul'. These forms enable plants to move themselves, something that non-living substances cannot do.[33] Next in line come the souls or substantial forms of non-human animals, which have emergent properties to an even greater degree than the souls of plants, since in virtue of them non-human animals can cognize the world. In Aquinas's view, although such operations require bodily organs, they have nothing at all to do with the qualities of the elements (again, the most basic features of corporeal matter).[34] Finally, the substantial forms of human beings have operations that do not require bodily organs at all in order to operate, although they are designed to work in tandem with bodily organs.[35]

As we have seen, some substances have a special name for their substantial forms; the substantial forms of living substances are called 'souls'.[36] Thus, some substantial forms are souls and some are not. But some souls are such that they are able to survive being separated from matter. Human souls are such substantial forms and, unlike all other substantial forms of material objects, they have operations, namely understanding and willing, that rise completely above the nature of matter, and so these substantial forms can survive apart from the matter that they naturally configure. In contrast, the substantial forms of non-human material substances are immersed in matter such that they go out of existence whenever they are separated from it.[37] Following B. Bazan, we might refer to human souls as 'subsistent substantial forms'.[38] Aquinas himself sometimes calls the substantial forms of non-human material substances 'material forms' (*formae materiales*)[39] or 'natural forms' (*formae naturales*).[40]

Since material substantial forms lose their existence whenever they are separated from matter, it stands to reason that such substantial forms have their origin in matter as well. Indeed, this is Aquinas's view. Aquinas also thinks that material substantial forms are not created by God.[41] Material substantial forms originate by way of a natural process, and not a miraculous one. But neither are material substantial forms *generated*.[42] This is because only beings, properly speaking, are generated according to Aquinas; and, as we have seen, material substantial forms are not beings but principles of being.[43] Rather than being generated, Aquinas thinks that material substantial forms are 'educed' (*educitur*) from matter.[44] For now we can note that by 'eduction' Aquinas means the drawing out of a substantial form from matter that is 'in potency' to a way of substantial being; substantial forms that are not currently configuring some matter *can* come to configure some matter through the actions of agents extrinsic to that matter manipulating it in various ways.

In contrast to material substantial forms, the human soul is able to exist apart from the matter it configures. The soul is thus a subsistent thing for Aquinas, and not simply a principle of being as are material substantial forms.[45] The soul's subsistence is tied to its having operations that transcend matter in the relevant sense: the distinctive operations of the human soul function apart from matter, and, with God's help, the human soul as a whole *can* function apart from matter.[46] However, even when it is separated from matter, a human soul remains the substantial form of a human being. Although the human soul can exist apart from matter, existing separately from matter is an unnatural state for the soul. Finally, since human souls are subsistent entities, they cannot have their origin in matter.[47] Thus, unlike material substantial forms, human souls only exist by way of a special act of creation on the part of God.[48]

Aquinas's views on the human soul raise a puzzle. Given Aquinas's view that human souls are not only subsistent entities but beings, he would seem to be committed to the following disjunction:

(HB) (1) A human being is identical to a human soul (the body is really outside the species 'human being' altogether) or (2) a human being is one substance composed of two actual beings, namely soul and body.

Aquinas thinks HB is false. But is such a view consistent with what he says about the human soul being subsistent?

Aquinas does not deny that the human soul is a being. The soul does indeed have its own existence for Aquinas. However, Aquinas denies that the human body has its own existence in addition to that of the soul. Rather, Aquinas says that the human soul 'communicates' its existence to the matter that it configures.[49] Thus, a human being is one thing composed of two metaphysical parts: a metaphysical part that has its own existence, the soul, and some matter, which shares the soul's existence when it is configured by it. Thus, although the human soul is a being, a human being is not a single substance composed of many substances. Aquinas is right to reject part 2 of HB.

But now it should be clear why Aquinas is right to reject part 1 of HB as well. As we have seen, the soul is only a metaphysical part of a human being,[50] and so a human being is not identical to her soul.[51]

Of course, Aquinas does think that Socrates's soul remains in existence upon Socrates's biological death. Furthermore, Aquinas also thinks that it is correct to refer to the separated soul of Socrates as 'Socrates'.[52] It thus appears as though Aquinas's views are inconsistent where the relation between a human being and her soul is concerned. But this is only if we take 'is' necessarily to signify the identity relation in this context. 'Socrates is his soul' can be taken in two senses: (a) Socrates is *identical* to his soul, and (b) Socrates is *composed of* his soul. As we've seen, Aquinas thinks (a) is false. Socrates is essentially a human being; human beings are rational animals; animals are material beings, and souls are not material beings.

But Aquinas thinks that Socrates is sometimes just his soul, namely when Socrates is separated from his body at death. In point of fact, human beings, because they are animals, are usually composed of a substantial form and some matter. This is the natural state of affairs for any human being. However, since the human soul is that part of a human being that is by itself sufficient to preserve the identity of that human being, and the human soul is able to exist apart from matter, sometimes a human being is composed of only her soul, and not of her soul and matter, the two metaphysical components that she naturally has in virtue of being a human being. Thus, although Aquinas rejects that a human being is *identical* to her soul, he does think that a human being can be *composed* of only her soul. Just as with a substance's integral parts, so also with the metaphysical parts of a human being: composition is not identity.

We might give Aquinas's views with respect to the metaphysical parts of a human being the following formulation (call it the 'normal' or 'natural' condition of a human being where the metaphysical parts of a human being are concerned, 'NC' for short):

(NC) A human being is normally (or naturally) composed of a substantial form (the soul) and some matter (the body).

However, since the human soul is by nature something subsistent, and the substantial form of a substance x is sufficient to preserve x's identity through time and change, it is possible that

(PC) A human being is composed of her substantial form alone.

Because composition is not identity for Aquinas, however, PC is compatible with Aquinas's view that human beings are not identical to their souls.[53]

As Eleonore Stump points out, this view of Aquinas's receives some support from arguments made by contemporary philosophers based on some plausible-sounding intuitions about human identity.[54] We might think a human being is essentially an organism. We might also think that the existence of the brain (or some part of the brain) of the human organism is a necessary and sufficient condition for the existence of a human organism. Imagine that it were possible to keep a human being named 'Jane' alive after a terrible accident when all that remains of Jane after an accident is her head. Since Jane is essentially a human organism and not a human head, it is not the case even after the accident that Jane is identical to her head. Nonetheless, as we are imagining, a head (or brain, or brain part) is a part of a human organsim that is sufficient to preserve a human being's numerical identity over time and change.[55] Jane is not identical to her head but she might be composed of just her head. Although this view deals with what Aquinas would call the 'integral' parts of a human being, and not her metaphysical parts, the view is at least analogous to Aquinas's view that a human being is possibly composed of, but never identical to, her soul.

Finally, it should be mentioned that the state of being separated from matter is an abnormal or unnatural one for the human soul. The soul, by its very nature, is a substantial form of a material substance.[56] Thus, the life enjoyed by a human being composed of only her soul is incomplete and, given Aquinas's belief in a good and loving God, he thinks such a state can only be a temporary one.[57]

To summarize Aquinas's views with respect to human substantial forms, like a material substantial form a human substantial form is not a substance for Aquinas, but always remains a metaphysical part of a substance. But, unlike a material substantial form, a human substantial form can survive being separated from the matter it normally or naturally configures. According to Aquinas, this is because the human soul has operations that transcend matter and do not require a body. Nevertheless, since the separated soul is not a complete substance, but only part of a substance, and because it naturally makes use of matter in order to perform its distinctive operations of understanding and willing, the state of being separated from matter is an abnormal or unnatural one for the human soul. Finally, Aquinas thinks that, although the human soul is not identical to a human being, a human being sometimes is composed of only her soul, namely after death. These views of Aquinas's provide reasons for thinking that he would be very sympathetic to those contemporary philosophers who argue, 'constitution is not identity'.

Aquinas on prime matter

Like the term 'form' for Aquinas, 'matter' is a term used with a variety of related, yet different meanings. For example, Aquinas uses 'matter' to refer both to entities that are parts of substances and entities that are substances.[58] But the sense of 'matter' at the centre of Aquinas's metaphysic of material objects is that of 'prime matter' – the locution Aquinas uses to refer to that wholly non-formal recipient of the substantial form of a material substance.

Aquinas often explains what he has in mind by prime matter by first comparing it to a substance insofar as that substance functions as the subject of the accidents that modify it.[59] So for example, in *De principiis naturae*, Aquinas talks about the nature of prime matter by comparing it to the matter of a statue, which is a substance for Aquinas, e.g. some bronze.[60] How is some bronze like prime matter? They both function as the matter for some form, or to put it another way, they both function as that which a form configures. Although Aquinas thinks that there are some forms that do not configure anything at all, namely, immaterial substances,[61] all material substances have a part that is configured by form. Thus, one way to think of the meaning of 'matter' for Aquinas is as that part of a material substance that is configured by form. However, in the case of material substances (in contrast to accidental beings such as artefacts), the matter that is configured by form is not *in and of itself* an existing subject. Prime matter only exists insofar as it is being configured by a substantial form. Thus, it is a subject only in the sense that it receives a form. This recalls what I said about typical substantial forms earlier: they are not beings, but principles of being. The same is true of prime matter. Prime matter does not exist without a corresponding substantial form.

In the passage from DPN we have been looking at Aquinas mentions prime matter in the context of discussing the metaphysical composition of a material substance *at* a time. The assumption seems to be that all material objects must have a first matter, that is, something that gets configured but does not have configuration itself. More often, however, Aquinas posits the existence of prime matter in attempting to make sense of certain radical changes we observe taking place in the world.[62] In order for something genuinely to change, Aquinas thinks three things must be the case. First, there must be some subject s of the change in question, that is something that remains numerically identical through the change. Second, there must be a privation of some feature f in s prior to the change. Finally, there must be some feature f in s subsequent to the change. Now it is clear, Aquinas thinks, what sorts of things stand in for the variables above in a case that involves some material substance undergoing a change in quantity. For example, a dog (s) lacks the feature of being two feet tall at the shoulder (not-f) at time t but comes to have the feature of being two feet tall at the shoulder (f) at $t+1$. But what about a case where the change in question is so radical that something goes out of existence (or comes into existence)? Take the death of a living organism as an example.[63] The living organism itself cannot be the subject for such a change, since the living organism does not survive the change in question (a change by definition involves one and the same thing *remaining* through the change, that is, through the loss and gain of properties). Furthermore, none of the integral parts of the organism (nor the mere sum of these integral parts) could serve as the subject for such a change according to Aquinas, since the integral parts of a substance x exist only in virtue of x's substantial form, and x's substantial form no longer configures matter at the point in which x is corrupted. Since the death of an organism is a *substantial* and not merely an *accidental* change, this leaves as the only remaining possibility that the subject of substantial form is not itself a substance but something that in and of itself is in potency to substantial existence. This is what Aquinas calls 'prime matter'.

Why must there be a subject that remains numerically identical through every change? Well, if there were not a subject that remains through a change, what we would have in such a case is something miraculous: at one moment there is a subject *s* at a place *p*, while at the next moment *s* is annihilated and a subject completely different from *s* is created *ex nihilo* at *p*. But it is surely absurd to imagine that this is what happens every time something is generated or corrupted in the natural world. (Thus, *x*'s being created needs to be contrasted with *x*'s being generated. The former involves the miraculous bringing of *x* into existence *ex nihilo*, whereas the latter is the natural process whereby *x* comes into existence out of (from) matter that pre-exists the generation of *x*, where the matter out of which *x* is generated is configured by a substantial form other than that of *x*. Likewise, the annihilation of *x* – the reduction of *x ad nihilum* – should be contrasted with *x*'s corruption, where *x*'s matter comes to be configured by a substantial form other than that of *x*.)

Perhaps there is another way to make sense of radical changes such as the death of an organism: reject the idea that an organism goes out of existence when it dies; the very same organism can have the property *being alive* at one moment, and at the next moment, the property *being dead*. The problem with this interpretation of radical change is that 'organism' means 'a living substance'; thus 'dead organism' is really a contradiction in terms. But why not say that it is a body that has the property *being alive* at one moment, and *being dead* at the next?[64] After all, bodies come in living and non-living varieties. However, for Aquinas, this would entail that such features of the body were accidental forms. This is because an accidental form is one that comes to an already existing substance and modifies it. If the very same body (where 'body' picks out a substance of some kind) can be alive or dead, that means that those features are accidents of that body. Aquinas would therefore reject this interpretation of death on the grounds that it construes a living thing to be an accidental being (a combination of a substance (the body) and an accident (being alive)). But living things are substances *par excellence* for Aquinas. A modern-day reductionist may have no problem construing living things in such reductionistic terms, but not Aquinas.

But what then is the subject of such a substantial change? Given Aquinas's view that substances are not composed of actually existing substances or, alternatively, that every material substance has only one substantial form,[65] the ultimate subject of change must be something lacking form altogether.[66] This is, of course, prime matter. Aquinas thinks that prime matter has the character of being formless in and of itself, but that prime matter never exists without some substantial form configuring it. We can thus think about prime matter from two perspectives: we can consider prime matter in and of itself apart from its being configured by some substantial form, and we can consider it insofar as it exists as a metaphysical part of a material substance, that is insofar as it is configured by a substantial form.[67] Let me say something about prime matter from each of these two perspectives.

Aquinas calls prime matter 'prime' because it does not have any matter prior to it.[68] In other words, prime matter is matter that itself has no material component. It is literally the 'first' matter. Of course, one might endorse the possibility of an 'Oriental Boxes theory' of matter,[69] namely, the view that says that all matter is itself composed of matter and form (to put it another way, every material object belonging to some stuff-kind K, e.g. some bronze, is composed of some stuff not-K). Aquinas rejects this view of matter on account of the fact that it would make the explanation of radical change impossible: if all matter is itself a composite of form and matter, then there is no single subject that itself remains through radical change – as we've seen, a necessary condition for making sense of any event that is reasonably termed a 'change' for Aquinas.[70]

Aquinas also talks about the nature of prime matter in and of itself in another way. As he puts it, prime matter is by its nature 'in potency' to substantial form,[71] or substantial being.[72] What does Aquinas mean by the phrase 'in potency' in these contexts? Passages in DPN might give one the impression that what Aquinas means by 'x is in potency to y' is just that 'if x is in potency to y, then x can be y, but x is currently not y'.[73] But such a *general* expression of the meaning of 'x is in potency to y' is ambiguous for Aquinas. This is because there is an important difference between the way that prime matter is in potency to form and the way that substances are in potency to form. As Aquinas explains:

> Properly speaking, what is in potency to substantial being is called 'prime matter'; but what is in potency to accidental being is called 'a subject'. Hence, it is said that accidents are in a subject; however, it is not said that a substantial form is in a subject. And in accordance with this, matter differs from a subject, since a subject does not have being from that which comes to it, but has complete being *per se*; just as a human being does not have being from whiteness. But matter has being from that which comes to it, since with respect to itself it has incomplete being.[74]

Thus, to make sense of 'prime matter is in potency to y' one must keep in mind that *through itself* prime matter does not actually exist; prime matter has actual existence only because of the substantial form that configures it.[75] Thus, for prime matter to be in potency to y would appear to mean that prime matter can be a part of a substance y, but it is currently not a part of a substance y, and since prime matter does not exist on its own, it must now be part of a substance not-y.[76] Although prime matter is thus like any material substantial form in that it is a principle of being and not itself a being, prime matter differs from substantial form in that prime matter explains those radical changes in nature that involve the generation and corruption of material substances whereas substantial form explains why matter (in *this* rather than *that* place) comes to be some particular substance belonging to a particular species.

Finally, because the subject of radical change cannot itself have any substantial forms (nor can it have any accidental forms, since accidental forms modify actually existing beings, namely substances), Aquinas thus defends the doctrine that prime matter, in and of itself, is pure potentiality for substantial form.[77] In other words, prime matter itself is something purely non-formal.

Having examined prime matter from the perspective of its nature in and of itself, let me say just a few things about prime matter insofar as it actually exists. Insofar as prime matter is a metaphysical part of a material substance, it is not simply an abstraction, a purely conceptual being, or a being of reason. For surely a mere abstraction could not function as the subject for a real change. Nevertheless, insofar as prime matter actually exists, it does so only as one of the metaphysical parts of a material substance, and thus as a principle of being and not as a being itself. Prime matter, considered apart from any substantial form giving it actual existence, is a mere abstraction.

Recall also that, for Aquinas, all actual material substances are beings that are spread out in three dimensions. This suggests that whenever prime matter is made actual by substantial form, the substance that has prime matter as a component is something spread out in three dimensions. That is to say, prime matter always exists as what Aquinas refers to as 'designated matter' – matter that can be pointed at with the finger.

Aquinas's doctrine of the unicity of substantial form

Aquinas argues throughout his career that every substance has only one substantial form.[78] Aquinas has several different reasons for holding the view. But before we look at these, let us first see what the 'unicity doctrine' amounts to.

If a substance has only one substantial form, this means that all of a substance's essential features are actually present in that substance because of that substance's substantial form. For example, the substantial form of a human being causes it to be a being, a substance, a corporeal thing, a living thing, an animal and a rational being. A material substance's one substantial form also causes it to be spread out in three dimensions and to have the different kinds of integral parts that it does. Thus, the unicity doctrine has a bearing on Aquinas's understanding of the integral parts of a substance. Indeed, we can go so far to say that the integral parts of a substance are not substances for Aquinas because of his doctrine of the unicity of substantial form: since integral parts do not have their own substantial forms giving them existence, but rather have existence from the substantial form of the whole – and only things with substantial forms are substances – it follows that the integral parts of a substance are not themselves substances. Aquinas's doctrine of the unicity of substantial form is thus closely connected with a view that I shall investigate further in discussing Aquinas's understanding of the integral parts of substances: that substances are not composed of substances.

Aquinas's views on substantial forms also entail that no form of a substance survives the generation/corruption of that substance. For example, Lynne Rudder Baker has the intuition that a person's body survives the death of the person constituted by that body.[79] So, after a person dies (call her 'Jane'), it is still literally true that the human-shaped thing in the coffin is a human body; indeed, it is numerically identical to the one that constituted Jane before her death. As we saw above, Aquinas rejects this interpretation of what happens at death. In his view, the human-shaped thing in the coffin is only a human body in an equivocal sense. He compares the human-shaped thing in the coffin to a statue of a human being, which is certainly not literally a human being at all. Furthermore, this goes for the parts of a substance as well. The eye-shaped thing in a corpse only looks like a real eye; in fact, it is only an eye in an equivocal sense.[80] More important for our purposes is why Aquinas would reject Baker's intuition. And the answer has to do with the doctrine of the unicity of substantial form. Because it was the presence of a human substantial form that explained why Jane's body was a human body, when Jane's substantial form departs from its material conditions, so do all the features caused by that substantial form, including the feature *being a human body*. However, what looks like Jane's body can still be a *corporeal thing* after Jane's substantial form departs. This is because human bodies are living material substances for Aquinas, but not all corporeal things (bodies) are living things.

So what is the human-shaped thing in the coffin that resembles Jane's body if it is not a human body according to Aquinas? Although Aquinas never answers this question explicitly to my knowledge, he could accept one of the following two answers. The human-shaped thing that resembles Jane's body is either a substance belonging to some strange species, 'corpse', which would be a kind of compound for him, that is, a non-living material substance, or perhaps more plausibly, the human-shaped thing is not a substance at all, but rather an aggregate of substances, namely, a collection of elements and compounds belonging to those material kinds that go into making up the parts of living substances.

Notice a further implication of the doctrine of the unicity of substantial forms: no accidental features survive the generation/corruption of a material substance. Take again the example of Jane's body and the human-shaped body in the coffin. Assume for the sake of argument that Jane's body and the body in the coffin appear to be indistinguishable as far as their shapes are concerned: the body in the coffin looks just like Jane's body. But for Aquinas the body in the coffin's feature *being in the shape of Jane's body* might be the same in *kind* as Jane's bodily feature *being in the shape of Jane's body*, but it could not be the same in *number*. This is because for Aquinas accidents (of which *being in the shape of Jane's body* is a perfect example) are individuated by the substances those accidents modify, and Jane and the body in the coffin are two numerically different substances.

What goes for accidents having to do with shape goes for other accidents as well. If substance x is corrupted at time t so that a substance other than x (call it 'y') comes into existence, then none of y's accidents will be numerically identical to accidents had by x, even though many of the accidents had by x and y are the same in kind. This follows from Aquinas's unicity doctrine and his view that accidents are individuated by the substances they modify: substances have only one substantial form, and so x and y don't have any integral parts in common that might serve as a substratum for the preservation of a numerically identical accident. Since accidents are individuated by substances, if x and y are numerically different substances by definition, then x and y don't have any numerically identical accidental features.

This brings us to Aquinas's reasons for holding the doctrine of the unicity of substantial forms. One way Aquinas argues for the unicity doctrine centres on his understanding of the nature of substantial and accidental forms.[81]

> A substantial form differs from an accidental form by the fact that an accidental form does not make [something] to be *simpliciter*, but to be such, just as heat does not make a subject itself to be *simpliciter*, but to be hot. And therefore, upon the coming of an accidental form something is not said to be made or generated *simpliciter*, but to be made such or to be related in some way. Similarly, when an accidental form is lost, something is not said to be corrupted *simpliciter*, but in a certain respect. However, a substantial form makes [something] to be *simpliciter*, and therefore through its coming something is said to be generated *simpliciter*, and through its receding to be corrupted *simpliciter*. . . . If therefore it were such that besides the intellectual soul there pre-existed in matter some other substantial form through which the subject of the soul were an actual being, it would follow that the soul does not make [something] to be *simpliciter*; and consequently that it is not a substantial form; and that through the coming of the soul there would not be a generation *simpliciter*, and neither through its leaving corruption *simpliciter*, but only [a generation and corruption] in a certain respect, which are [both] manifestly false.[82]

As Aquinas notes, substantial forms make a thing to be a substance – something capable of having accidental features – while accidental forms cause a pre-existing substance to have accidental features such as certain qualities, quantities, relations, etc. If a form f were to come to something that already had a substantial form, then f would have to be an accidental form. Thus, a substance cannot ever have more than one substantial form.

Notice also the way that Aquinas's unicity doctrine ties into his understanding of prime matter. Matter that is prime must take on a substantial form because it cannot take on an accidental form; accidents only modify substances. Furthermore, the composition of prime matter and a substantial form is a substance. But forms that modify substances are accidental forms. Thus, if prime matter is already configured by a substantial form, it cannot then be configured by yet another substantial form. Thus, every substance has only one substantial form.

Another reason that Aquinas gives for the unicity doctrine is that a substance is one thing in the most unqualified way, and a substance's having more than one substantial form is incompatible with such a view. The following is a representative argument of this type from the SCG and, although it argues specifically that human beings cannot have more than one soul, the argument also shows that any substance can have only one substantial form.

> Something has being and unity by the same thing, for unity is consequent on being. Therefore, since every single thing has being by form, it will also have unity by form. If therefore many souls as diverse forms are posited in a human being, then a human being will not be one being, but many. Neither will an order among forms suffice [to give] unity to a human being. This is because to be one according to order is not to be one *simpliciter*, since unity of order is the least of unities.[83]

This argument explicitly appeals to the convertibility of being and unity, and the view that forms cause being. However, what is assumed in the argument is that forms are either substantial or accidental, and that substantial forms cause substantial being, whereas accidental forms cause beings that already exist to be modified in a certain way. If a form comes to something that already exists as a substance, then such a form is an accidental form, e.g. to be one according to order as in a heap of stones, and such a form makes merely for an accidental being. But substances are not mere collections or heaps of substances; rather, they are unities of the highest order. Therefore, each substance has only one substantial form that causes it to be a being, a substance, and the fundamentally unified kind of thing that it is, e.g. a dog.

Aquinas's defence of the unicity doctrine thus depends primarily upon his understanding of the nature of substantial forms, accidental forms and prime matter. Many of Aquinas's contemporaries parted company with Aquinas on the matter of the unicity doctrine, but they did so because, for one reason or another, they thought about the nature of substantial forms and prime matter in ways that differed from that of Aquinas.[84] However, disagreements with Aquinas on the nature of form and matter need not detain us here. What can be said is that Aquinas's interpretations of the doctrines of substantial form and prime matter are coherent. Furthermore, we can note what Aquinas's tenacious adherence to the unicity doctrine tells us about his own view of substance. Material substances are those composed beings that are unified in the highest possible degree. In and of itself, every material substance has only *one* fundamental configuration, that is one substantial form that causes it to be a substance as well as to be the fundamental kind of thing that it is.

Aquinas on the integral parts of material substances

In addition to the metaphysical parts of material substances, Aquinas also has much to say about the ordinary parts of material substances such as hands, heads and hearts. Before explaining Aquinas's views on the ontological status of these parts, I need first to make some remarks regarding terminology.

Aquinas sometimes refers to the ordinary quantitative parts of a substance such as hands and hearts as 'parts of a substance'.[85] In other places, Aquinas refers to the ordinary parts of things as integral parts (*partes integrales*).[86] An integral part is any part that is related to its whole in such a way that the whole is not present to that part by its entire essence or power,[87] or alternatively, an integral part is a part that is related to a whole such that the whole cannot be predicated of that part.[88] By this last description it should be clear that Aquinas does not always use 'integral part' and 'ordinary physical part' as equivalent expressions since the metaphysical parts of substances are neither ordinary physical parts, nor can they be predicated of the wholes to which they belong. However, for the sake of convenience and because Aquinas does often employ 'integral part' to refer to the ordinary quantitative parts of material substances, I will hereafter refer to the ordinary quantitative parts of a material substance as that substance's 'integral parts'.

There appear to be at least two broad kinds of integral parts for Aquinas: the integral parts of elements and compounds on the one hand, and the integral parts of heterogeneous substances such as the higher animals on the other.[89] Before I speak about how these kinds of integral parts differ from one another, I want first to say something about integral parts in general.

Every integral part of a material substance is something such that it has smaller dimensions than the whole to which it belongs.[90] The integral parts and the metaphysical parts of a substance can be contrasted on this score, since a substance's metaphysical parts do exist in some sense wherever the substance as a whole exists.[91] Furthermore, those objects that are excluded from the range of causal influence of the substantial form and matter of substance *x* are not integral parts of *x*. In contrast to a substance's metaphysical parts, an integral part always exists as a sub-portion of a substance.

Like metaphysical parts, however, integral parts are not substances according to Aquinas.[92] Although his descriptions of integral parts as compared to substances vary a bit throughout his corpus, there does appear to be a uniformity of doctrine on the matter. For example, in one place he says that integral parts of a substance 'are not called particular substances, which, as it were, are subsistent things *per se*. But they [i.e. parts of substances] subsist in the whole.'[93] Here integral parts do subsist – thus, an integral part is a *hoc aliquid* – but integral parts do not subsist in themselves but only in virtue of the whole of which they are parts. In another passage, Aquinas remarks concerning integral parts, 'although they do not exist in another as a subject, they do not exist in themselves, but in the whole'.[94] (Recall that to be something such that it does not exist in another as a subject is to be a *hoc aliquid*, i.e. something subsistent.) Aquinas does ascribe subsisting *per se* to integral parts in one place, but denies that integral parts are complete substances since they are incomplete in their species.[95]

Thus, Aquinas has the overall view that integral parts exist *in* another thing, although they differ from accidents in this regard since integral parts do not exist in another as a subject.[96] Integral parts are therefore unlike accidents in being subsistent entities, but like accidents in that they do not have a complete essence or species.[97]

Aquinas has a number of ways of making this point about an integral part's incompleteness in species. Sometimes he denies that the integral parts of a substance belong to a species at all, properly speaking. It is rather substances that belong to a species.[98] However, more often he argues that integral parts belong to a species only in virtue of being parts of a whole that itself has a species *per se*, that is a substance.[99]

Why do integral parts such as hands and hearts not have a complete species? Aquinas seems to have at least two reasons for thinking about integral parts in this way. The first argument is based on what Aquinas takes to be some relevant empirical evidence: integral parts such as hands and feet no longer function when the whole to which they belong is corrupted. Because something's existence and being what it is are tied to that thing's distinctive operations – if something cannot perform its specific operations it no longer is a thing of that kind – integral parts have their species only as long as they are functional parts of the wholes to which they belong. As Aquinas notes in many places, the integral parts of living organisms that we call 'eyes', 'hands', 'flesh' and 'bones' are called by such names only equivocally after the departure of the soul from the body; they are as much the integral parts of an animal at that point as are the 'eyes' of an animal portrayed in a painting or sculpture.[100] In one place, Aquinas sums up this interpretation of the evidence by way of the following axiomatic statement: 'No part has the perfection [or completion] of a nature [when it is] separated from the whole.'[101]

Given Aquinas's position that the integral parts of a substance are what they are because of the substance to which they belong, it would seem that we can argue in reverse that severed limbs are also integral parts only in an equivocal sense. Indeed, a limb such as a hand that has been separated from a living human being is no longer under the guidance of that human being and so is no longer made something actual by that human being's soul. A severed limb should therefore demonstrate a lack of its characteristic function. And this seems to be the case. A severed hand cannot grab, scratch, hold on to something, etc. Thus, it is no longer really a hand.[102]

Aquinas has a second argument to support his view that the integral parts of a substance have their species in virtue of the wholes to which they belong. Aquinas argues that the substantial form of a substance gives form (species) and existence not only to the substance as a whole, but its parts as well. As he observes, a form that comes to a being already having a substantial form is an accidental form.[103] If every integral part of a human being had its own species (because it had its own substantial form), then the form of the whole in a human being would stand in relation to its integral parts as an accidental form. But this would mean that the human form is merely an accidental and not a substantial form. But the human form is substantial. Thus, integral parts are not themselves complete in species. The integral parts of a substance must exist and be what they are in virtue of the substance to which they belong, otherwise the human being is reduced to being a mere aggregate of substances.

In addition to saying that the integral parts of substances are not complete in species, Aquinas has another way of explaining why integral parts are not substances. He sometimes says that parts of substances are not actual, but rather potential, substances.[104] In other words, while an integral part exists in a whole, it is not actually a substance, and therefore does not have a complete species, but when that part is released from its whole, it can become an actual substance complete in species (or alternatively, an aggregate of substances, each member of which is complete in its species). To put it yet another way: the releasing of an integral part of a substance is an occasion for the generation of a substance (or an aggregate of substances). For example, some water (a substance for Aquinas) has a part p (which is also some water) at time t. At $t+1$ p is released from the whole to which it belonged. At t, p was potentially a substance insofar as it was a part of a substance. But at $t+1$ p is an actual substance belonging to the kind, *water*. To take another example, Socrates has two hands at t. At t, Socrates's left hand is potentially an aggregate of substances, e.g. (for Aquinas) some water, some earth, etc. At $t+1$ Socrates dies and what was Socrates's hand at t and only potentially an aggregate of substances becomes an aggregate of actual substances at $t+1$.

But why should we think that an integral part is only potentially a substance and not actually a substance as long as it is a part of a substance? Aquinas has at least two reasons for this view. One of these falls back on a familiar theme: integral parts, insofar as they are parts of a whole, that is exist *in* a whole, are not things that subsist *per se*.[105] Recall the definition of substance I gave above: substances subsist *per se* and are complete in their respective species. But parts have their identity in virtue of being parts of the wholes to which they belong. Thus, at least while they are parts, they do not have a complete species apart from the whole; integral parts are not actual substances.

A second reason integral parts cannot be actual substances as long as they are parts of a substance is as follows: if they were, then substances could be composed of substances. But substances *cannot* be composed of substances.[106] Therefore, the integral parts of substances, e.g. a hand, or a quantitative part of some portion of water, cannot be substances as long as they are parts of those substances.

What are Aquinas's reasons for holding the view that substances cannot be composed of substances? As Aquinas puts it succinctly in one place from his commentary on the *Metaphysics*: 'two actual things are never one actual thing; but [there are cases] where two things that are in potency are one actual thing, as is clear in the case of the parts of a continuous thing.'[107] For Aquinas, substances are unified things of the highest order. If the integral parts of substances were actual substances, those parts would enjoy a unity greater than the whole of which they were parts. Of course, Aquinas admits that there are entities that have substances as parts – but such entities are not substances themselves. Thus, it is rather the case that the parts of substances are potential substances as long as they are parts, and not actual ones.

Later in the same passage, Aquinas argues:

> Every thing is separated from another through its own form. Thus, by the fact that some things become one actual thing, it must be that they all are included under one form, and not that each should have a single form through which each of them would be actual. Hence it is clear that, if a particular substance is one, it will not [be composed] of substances actually existing in it.[108]

If the parts of a substance x were substances, then these parts would have their own substantial forms that separated them from each other and from x. But this would be to destroy the unity of x, such that x couldn't itself be a substance. If the parts of x were substances, then x's form of the whole would be an accidental form only. But x is a substance. Therefore, the integral parts of x are not actual substances.

A sceptic might remark that Aquinas is certainly right to say that two things cannot be one thing, and therefore so much the worse for there being any compound entities. In this sceptic's mind, the only real material things are simples: atoms, quanta or what have you. Aquinas is aware of a position like this; in the very same passage I have been citing he attributes a view like this one to

Democritus. But Aquinas takes the crucial error of Democritus to be his failing to distinguish between actual and potential being.[109] As we shall see in the sequel, Aquinas's view that integral parts are potential substances has some interesting implications for debates about material constitution.

We are now in a position to offer a definition of 'integral part of a substance'. I propose the following:

(IP) x is an integral part of a substance $y =_{df} x$ is a part of y such that (1) x is a *hoc aliquid*, (2) x is incomplete in species, both in the sense that (i) x is what it is in virtue of being a part of y, (ii) x does not exist apart from the integral parts of y other than x, and (3) x has dimensions smaller than those of y.

The definition distinguishes the integral parts of substances from actual substances, accidents and the metaphysical parts of a substance. Clause 1 distinguishes integral parts from accidents and non-subsistent metaphysical parts. In contrast to Socrates's matter (a non-subsistent metaphysical part) and Socrates's whiteness (an accident), Socrates's hand (an integral part) is a *hoc aliquid* – a being that does not exist in another as a subject. On the other hand, clause (2)(i) distinguishes Socrates's hand from things such as Socrates himself. Integral parts differ from substances on account of the fact that the latter but not the former are complete in species. An integral part is defined by way of reference to the substantial whole to which it belongs. A substance is defined through itself, and not in virtue of anything extrinsic to it. Finally, clauses (2)(ii) and 3 distinguish integral parts from subsistent substantial forms (human souls). Although integral parts subsist – they do not exist in another as a subject as do accidents – integral parts do depend upon the whole to which they belong for their existence (and their species). In contrast, subsistent substantial forms subsist *per se* and not *per aliud*; they can exist even apart from the whole to which they belong, that is subsistent substantial forms can exist apart from the form/matter composites to which they normally belong as parts. Furthermore, subsistent substantial forms exist in some sense just where the whole body does for Aquinas, and not, as integral parts do, in only a subregion of the body.

Having laid out Aquinas's understanding of the integral parts of substances, I want to say something about an interesting distinction that can be made between two different kinds of integral parts, a distinction that will be particularly relevant when we turn to discussing Thomistic solutions to the classical and contemporary puzzles about material objects. Compare material substances such as the higher animals, e.g. insects, reptiles, mammals, with portions of non-living substances, e.g. a puddle of water, a lump of bronze, etc. The former are heterogeneous substances – substances that have many different kinds of parts – whereas the latter are non-heterogeneous substances, every integral part of which belongs to the same kind as the whole. For example, split a higher animal into two and both parts will not be animals (although one of them may remain an animal). In contrast, split a portion of water in two and both portions will be portions of water.[110] Aquinas himself makes the distinction between these two kinds of integral part in a number of places.[111]

The integral parts of heterogeneous substances, parts such as hands, heads and hearts, might be referred to reasonably as 'functional' parts. This is because such integral parts are not simply quantitative parts of a whole. Each of them has a specific function that contributes to the distinctive activities of the material substance of which they are a part.[112] As Aquinas describes the parts of heterogeneous substances, although they are not actual substances, they are nonetheless entities 'close to actuality'. These integral parts are 'distinct according to form'.[113] However, functional integral parts are not substances since their functions are wholly directed towards the flourishing of the whole (organisms) of which they are the parts. For example, a hand has its own distinct operation, since it has a form that makes it distinct from other parts of the whole to which it belongs, but its operation, perhaps grasping objects, is not an activity done for its own sake, but for the sake of the organism of which it is a part.

However, not all integral parts have such distinctive operations. Some integral parts are merely quantitative parts of some whole. For example, take a quantitative part (call it 'x') of a portion of water (call the portion of water 'y'), e.g. the bottom third of the water that currently resides in Socrates's drinking glass. x is clearly an integral part of y since x is a quantitative subportion of y, and x is potentially a substance. Parts such as x might be called 'non-functional integral parts'.

But non-functional integral parts are funny sorts of parts. For one thing, a material substance that has them would presumably have an infinite number of such parts. Peter van Inwagen calls such parts 'arbitrary undetached parts' and thinks we have good reasons to think there are no such things, namely that they raise the PMC.[114] However, Aquinas certainly thinks that non-living substances have non-functional integral parts,[115] and so do some living things, namely those living things that are nearly non-heterogeneous such as worms and some plants.[116] Although he shows no signs of saying whether or not higher living organisms have such parts, I want to suggest that he need not be so committed.

Consider an arbitrary undetached part of Socrates, e.g. the left half of his body. Would Aquinas say that this is one of Socrates's integral parts? Given that he thinks non-heterogeneous substances have such parts, wouldn't it be ad hoc for him to say that some substances don't have such parts? I don't think so. Here's why. *Being heterogeneous* is an essential feature of some material substances (while *being non-heterogeneous* is an essential feature of some others), and whether or not a substance is heterogeneous has an effect on the kinds of parts it has. Non-heterogeneous substance-wholes, such as instances of the elements, have parts that are potentially substances while they are parts of such wholes. When these parts are released from their wholes, they become actual substances in their own right, not to mention substances that belong to the same species as the whole from which they were separated. In contrast, the integral parts of higher living organisms are all functional. Slice off a portion of Jane's skin from her left arm. It might appear as though this integral part is non-functional, that it is analogous to the bottom third of the water in Socrates's glass. But this piece of skin has a specific function: it protects a certain region of Jane's body from viruses, sharp objects and other foreign agents that might be harmful to Jane. As we shall see, higher living organisms do have some non-functional parts, but these turn out to be parts that are not properly understood as integral parts, namely a compound's elemental parts.

In closing this section I want to draw out a few implications of Aquinas's understanding of the integral parts of substances.[117] First, recall that in Chapter 3 I argued that Aquinas's views on the integral parts of substances have implications for his understanding of the extension of 'material substance'. We are now in a better place to see why this is the case. It appears as if there are *some* compound-kinds that do not have substances as instances. (Clearly some compound-kinds, e.g. bronze, do have substances as instances for Aquinas.) I have argued that Aquinas is committed to thinking that functional integral parts such as hands and feet cannot survive being separated from the influence of the substantial forms of the organisms to which they belong. A severed hand is not really a hand but rather an aggregate of compounds 'arranged hand-wise'.[118] But what goes for hands and hearts would seem to be the case also for things such as blood, bone and flesh. These things, as integral parts, exist only in virtue of the influence of an organism's substantial form. But blood, bones and flesh are compounds, that is, material entities that have elements as parts (see my discussion of elemental parts in the sequel). Thus, there are some compound-kinds that do not have substances as instances. Instead, all instances of compound-kinds such as blood and flesh are rather integral parts of organisms.

Second, Aquinas's views about integral parts also give him a way of preserving the reality of *compound* material objects without having to appeal to the counter-intuitive doctrine of spatial co-location: the view that says it is possible for two material objects belonging to different kinds to exist in the same place at the same time (this amounts to a denial of IMO5). Why do some philosophers espouse this counter-intuitive view? At least one reason has to do with the fact that there are cases of material constitution such that: (1) the constituted thing and the thing that constitutes it share all of the same parts, e.g. a statue and the clay that constitutes it or an organism and the aggregate of fundamental

particles that constitutes it, (2) the constituted thing and the thing that constitutes it have different essential properties, and (3) neither the constituted
thing nor the thing that constitutes it can be reduced to something non-
material, e.g. a property, event or process. A statue is a material object that
has essential properties that a piece of clay lacks, e.g. it can survive the loss of
a part, whereas a piece of clay is a material object that has essential properties
that a statue lacks, e.g., it can survive being flattened. In other words, co-loca-
tionists are essentialists who are committed to a non-reductionist (ontological
pluralist) approach to material reality. Since, for example, a statue and the
piece of clay that constitutes it are both material objects that share all of the
same parts, and the statue cannot be reduced to the clay, or vice versa, it is
possible that two material objects can exist in the same place at the same time.

Leaving aside the case involving artefacts (for now), let us focus on the case
involving an organism and the aggregate of particles that constitutes it. Aquinas himself accepts the common-sense intuition about material objects that
co-location is impossible.[119] But like the co-locationist, Aquinas believes (in
his own way) that things have essential properties and that there are compound material objects in the world, including such things as aggregates of
substances.[120] But Aquinas's view that the integral parts of substances are not
actual substances gives him good reason for denying that there is an aggregate
of fundamental particles present wherever any organism is present. The view
that every organism is constituted by an aggregate of fundamental particles
assumes that fundamental particles remain as actual substances whenever
they are parts of such organisms. However, as we have seen, Aquinas denies
that an organism's integral parts are actual substances, and so he is not committed to the view that an aggregate of actual substances constitutes an organism. Although integral parts and organisms are both real entities in Aquinas's
view, they each have a different ontological status: integral parts subsist only
in the wholes to which they belong, whereas substances are complete entities
in and of themselves. Because the integral parts of substances are not substances, there is no aggregate of substances spatially coincident with an organism for Aquinas. Thus, Aquinas need not commit himself to the possibility
of co-location in order to defend a non-reductionist approach to material
constitution.

Aquinas on elemental parts

In addition to composition out of metaphysical parts and integral parts, which
are modes of composition shared by all material substances, Aquinas thinks
that some material substances are also composed of what we might call 'elemental parts'. As we have seen, Aquinas takes compounds to be substances
that are composed of the elements. But non-living compounds themselves are

not heterogeneous substances. Compounds do not have quantitative (integral) parts that differ in kind from the wholes to which they belong, e.g. Aquinas thinks that every integral part of a piece of wood is some wood.[121] How can compounds be at the same time composed objects and non-heterogeneous? The answer lies in Aquinas's view that the elements compose substances in a way that differs from the way integral parts (and metaphysical parts) compose a substance.

Aquinas and his contemporaries observed that any compound *c*, when corrupted into its elemental parts, was thereby transformed into elements the same in species and quantity as those elements that originally were transformed into *c*. They concluded from such observations that elements somehow 'remain' in the compounds they compose.

One way to think about elemental presence in compounds is to say that elements *actually* remain in compounds, a compound being nothing more than a mere aggregate of substances every member of which belongs to some elemental-kind.[122] In other words, although compounds appear to be different in species from the elements that compose them, in reality they are not. We might call this a 'reductionist interpretation' of elemental presence today. None of Aquinas's contemporaries took this interpretation seriously, as it was clearly in conflict with experience and the teachings of Aristotle. Aquinas clearly rejects this view of compounds also.[123]

Aquinas views his own position on elemental presence in compounds as an explanation of a passage from Aristotle's *De generatione et corruptione*:[124] the substantial forms of the elements do not remain in a compound actually, but with respect to their power (*virtute*) only.[125] Aquinas gives his most detailed account of elemental presence by power in his treatise devoted to that topic, DME.[126] His idea here is that elemental bodies (substances) have certain *proper accidents* that are always present when such substances are present.[127] For example, some fire has the proper accidents, *being very hot* and *being very dry*. Because these proper accidents are *accidents*, they are forms that admit of being more or less. An object *x* might be hot to a certain degree, while another object is even hotter than *x*, and still another object is a bit cooler than *x*. When certain proportions of different elements enter into certain relations with one another, in the right circumstances the proper accidents of those elements are so changed that the elemental substantial forms are lost and a compound is generated that has a proper accident which itself is a mean between the extremes of the proper accidents of the composing elements. For example, say that bronze is composed of a certain proportion of fire and earth. Fire has *being hot* and *being dry* as its proper accidents, but it is *especially* hot. On the other hand, earth has *being dry* and *being cold* as its proper accidents, but it is *especially* dry. (I am imagining that) when fire and earth are brought together in the right proportions, relations and circumstances, the earth cools the fire, and the fire heats the earth such that a mean is reached, and this mean is a proper accident of bronze, and so now a substance that is bronze exists, but with earth and fire as its elemental parts. Thus, when this mean between hot and cold is reached, the substantial forms of fire and earth are lost and the substantial form of bronze is educed from the matter of the fire and earth. The elements of fire and earth remain insofar as a proper accident of bronze is a mean between the proper accidents of fire and earth, respectively.

Aquinas thinks this account of elemental presence successfully preserves the following facts: that compounds are substances and not just mere collections or aggregates of elements and that elements remain in the compounds they compose. Because the substantial forms of the elements are not actually present in the compound for Aquinas, a compound can be something more than just an aggregate of elements. And because the proper accidents of the elements remain in a compound – through the proper accidents of the compound – a compound really is *composed* of elements.

Aquinas sometimes talks about elemental presence in a slightly different way. In one passage from his commentary on the *Metaphysics*, Aquinas says that elements are present in a compound potentially and not actually.[128] Although Aquinas sometimes uses the phrase '*in potentia*' to refer to the power of a thing[129] – which suggests a connection with Aquinas's speaking about elemental presence in compounds *virtute* – in this passage Aquinas is focusing instead on the fact that the elements can be 'retrieved' from a compound when he says that elements are potentially in compounds.[130] Like the integral parts of a substance, elemental parts are potential and not actual substances as long as they remain elemental parts of a compound. For example, say we have a piece of bronze at time t. At $t + 1$ someone begins heating the bronze so that it is corrupted into its elemental parts, say some fire and some earth. At t, the fire and earth are present in the piece of bronze as potential but not actual substances. At $t + 1$, the fire and earth, no longer elemental parts of the piece of bronze, are now actual substances in their own right.

However, elemental parts and integral parts differ in the way that they are present in their respective wholes. Elements are present in compounds through their powers, the powers (proper accidents) of the elements being preserved in the proper accident of the compound they compose, where the proper accident of the compound is a mean power between the extremes of the proper accidents of the elements that compose that compound. In contrast, an integral part is present in a substance in virtue of being a quantitative part of that substance. Aquinas considers an integral part to be a *hoc aliquid*, although it enjoys this ontological status only relative to the whole to which it is a part. An elemental part is neither a quantitative part of a compound nor a *hoc aliquid*.

Christopher Decaen suggests that some interpreters of Aquinas (and Aristotle) on elemental presence consider elements to be integral parts, while others consider the relationship of elements to compounds to be a genetic one only, i.e. elements 'are ingredients in the mixture [compound] only in the sense that it came to be out of these elements and they will corrupt back into these elements'.[131] Given what we have seen Aquinas say about elemental presence, neither of these interpretations seems quite right. The 'genetic' interpretation argues that the elements are in no way parts of a compound. Instead, elements are simply the matter 'out of which' a compound is generated. But as we have seen, Aquinas thinks that elements, insofar as they exist in compounds, are parts of that compound.[132] Indeed, if parts were only of the integral variety, then these interpreters would be right to say that elements are in no way parts of compounds. But Aquinas also thinks that there are parts of substances other than a substance's integral parts. On the other hand, those who suggest that elements are integral parts also misconstrue Aquinas's understanding of elemental presence. As Aquinas makes clear, an elemental part of a compound is not located in a sub-region of that compound, as is the case with any integral part of that same compound. Now it may be the case that, if Aquinas were alive today, he *would* consider elemental parts to be a species of integral part, given current preferences for an atomistic interpretation of the nature of the fundamental physical entities. However, Aquinas's own position on elemental presence (which, of course, is a combination of metaphysical ideas *and* interpretations of empirical data) is that elemental parts and integral parts are two distinct genera of parts, and that compounds have both kinds of parts.

Let me close this section on Aquinas's account of elemental parts by comparing *elemental* and *metaphysical* parts. One way to see that elemental parts differ from metaphysical parts is by attending to the definition of an element. For Aquinas an element is a homeomerous substance: a substance x that is not composed of a kind of stuff that differs from x in kind. If a substance x were composed of *elemental* parts y and z, then x would not be an element but a compound. Thus, instances of elements such as *that* puddle of water do not have elemental parts.[133] But, according to Aquinas, there are no material substances that do not have metaphysical parts. All material substances have a substantial form and prime matter as component parts. Thus, elemental parts and metaphysical parts differ in their scope: not all material substances have elemental parts – instances of the elements by definition don't have elemental parts – but all material substances are composites of substantial form and prime matter.

But there is an interesting way in which elemental parts and metaphysical parts are similar: both kinds of parts constitute the essence of a material substance, albeit in different ways. We have seen that every material substance has a substantial form and prime matter as part of its essence. But elements too are included in the definition of all (terrestrial) substances. As we have seen, every material substance has only one substantial form, and this substantial form places that substance in its particular species. But the elements are present (by their power) in *all* non-elemental material substances as elemental parts. Aquinas thus says in his commentary on Boethius's *De trinitate* that composition out of the elements is included in the definition of such substances.[134] He contrasts the elemental parts of a human being on this score with a human being's integral parts, such as a finger, a foot or a hand. There are instances of human beings that lack these parts, but every (embodied) human being has the elements as parts.

To see why every human being has the elements as parts, recall that Aquinas takes human beings to be essentially rational animals. But animals are defined as sensitive, living, corruptible, material substances. Corruptible material substances are those made out of the four elements. Therefore, being composed of the elements is part of what it is to be an embodied human being.[135]

We have seen that Aquinas does not think that the different kinds of material composition – out of metaphysical, integral or elemental parts – are reducible one to another. The metaphysical parts of a thing are distinct from that thing's integral parts, which in turn are distinct from a thing's elemental parts (assuming the substance in question has elemental parts). Aquinas speaks about these three different kinds of parts in order to explain three different facets of compound material substances.

Aquinas on artefacts and composition

Aquinas is clearly an ontological pluralist. He believes that there are many different kinds of material objects in the world. As we've seen, however, not all material objects are material objects in the strongest sense according to Aquinas's way of seeing things. For example, the integral parts of material objects – objects such as hands, hearts and heads – are not actual substances as long as they exist as parts of substances.

Artefacts too are material objects that Aquinas does not regard as substances. Many contemporary philosophers would find such a view counterintuitive. According to these philosophers, artefacts are just as real as natural substances.[136] In this part I do two things. First, I offer some possible explanations for why Aquinas thinks that artefacts are not substances. Second, I explain Aquinas's own positive account of artefacts.

Artefacts are not substances

Aquinas denies that artefacts are substances throughout his corpus, sometimes only implicitly,[137] at other times, explicitly.[138] But why does Aquinas take this position?

One reason Aquinas thinks that artefacts are not substances is that he thinks that it is obvious that a material substance is not corrupted simply because it comes to be an integral part of an artefact. Imagine an artefact whose matter is an aggregate of material substances, e.g. an axe, which has as its matter a piece of wood and a piece of iron. For Aquinas, the axe is a composite of two material substances that serve as the matter for the axe and an accidental form that serves as the artefact's form of the whole (the accidental form in question perhaps being a relation between the two material substances, such as *being bonded in a certain way*). Thus, the piece of iron and the piece of wood retain their substantial forms while they function as the matter of the axe. If the axe were a substance for Aquinas, then the piece of wood that is part of the matter of the axe would be corrupted upon becoming a part of the axe. This is because, as we have seen, Aquinas thinks that a substance can't be composed of substances. But a piece of wood is not corrupted simply because it is fastened to a piece of iron – even if the wood-iron object is useful for some purpose. Therefore, the axe is not a substance. Aquinas makes an argument of this sort in DPN. Note that Aquinas thinks that a statue has an accidental form and not a substantial form because the statue's matter remains even though it takes on the form of a statue:

> When a statue is made from bronze, the bronze, which is in potency to the form of the statue, is the matter. The shape by which [something] is called a statue is the form. But [it is] not a substantial [form], since the bronze, before the coming of that form has actual being, and its being does not depend upon that shape. Rather, the form [of the statue] is an accidental one. For all artificial forms are accidental.[139]

Of course, one may respond that Aquinas gets it wrong here as far as the identity of the piece of bronze is concerned. The piece of bronze that exists prior to constituting any statue is not identical to the piece of bronze that constitutes the statue.[140] But this is certainly a counter-intuitive view. The piece of bronze, before it gets the statue shape, is identical to the piece of bronze that has the statue shape. Pieces of bronze are not corrupted simply because they take on a different shape. One might also respond here that Aquinas's views about substances and their integral parts – that the substances that precede the generation of a substance are not identical to any of the parts of the generated substance, since substances are not equal to non-substances – are highly counterintuitive in much the same way. But, as we have seen, Aquinas has good reasons for holding this view about the parts of substances, namely that substances cannot be composed of substances. An object composed of actual substances is not a unified object of the highest order, that is it is not a substance.

Aside from Aquinas's view that the substances that compose artefacts retain their identity whether or not they are composing an artefact, Aquinas appears to have at least one other reason for thinking that artefacts are not substances: artefacts are not natural things, and only natural things are substances.[141] Now Aquinas doesn't think that all natural things are substances; there are some naturally occurring things that Aquinas wouldn't count as substances, e.g. a pile of rocks or some muddy water. Thus, the proposal under consideration is that being a natural thing is a necessary but not a sufficient condition for being a material substance. Since artefacts would seem to be by definition unnatural things, artefacts are not substances.

One reason Aquinas thinks that substances must be natural things can be gathered by some remarks he makes about the essence of natural things in a passage from his commentary on the *Physics*. Unlike natural things such as substances, artefacts do not have a principle of motion or change *per se*. At the most, an artefact undergoes change because of the substances that it is composed of.[142] In contrast, substances have a principle of change *per se*. For example, living things grow (and thus change) because it is in their own nature to do so. Although living things depend upon other things to sustain their existence, we would not speak about things 'growing' if it were not for certain changes that originate from the living things themselves. More importantly for Aquinas's case here, he thinks that some *non-living* things also have a principle of change *per se*, e.g. fire moves to a higher place because of its nature, earth to a lower place, etc. To put it another way, Aquinas thinks that whatever exists *per se* has a distinctive operation *per se*.[143] Substances have distinctive operations *per se* because they exist *per se*.

Even if an artefact has a distinctive operation such as chopping wood, it has this operation only in virtue of the substances that compose it or *the attitudes of some agent extrinsic to the artefact*. For example, an axe is what it is because of the function that some human beings have given to it, namely, an axe consists of a piece of wood and a piece of metal bonded in such a way that the result is useful for chopping wood. Imagine that an axe is buried in the ground. A thousand years later, some people from a civilization different from the one that had originally constructed the axe uncover it. These people do not know what the axe *was* used for (perhaps they have no need for cutting wood, or for weapons), although they know it had *some* purpose or function (it has obviously been constructed by rational agents). The axe-finders begin to use the axe as a doorstop, for which it is quite useful. Does the axe thereby undergo a substantial change, changing from an axe (an artefact that is used to cut wood) to a doorstop (an artefact that props open doors)? We might think the answer is 'no', since something does not undergo a substantial change without undergoing any intrinsic changes.[144] But why then should we think that some wood and metal are thereby transformed into a substance simply because they have been fastened together in a way that is useful for chopping? Even if artefacts have distinctive operations or functions, they do not really have them *per se* as do natural substances, but only *per aliud*. Since substances have distinctive operations *per se*, artefacts are not substances.

Aquinas has another reason for thinking that substances must be natural things. Here Aquinas argues that, unlike substances, artefacts do not generate things of their own kind, that is they are not things that perpetuate their own species.[145] If an artefact were a substance, then it would be able to contribute to the generation of (or, itself be generated by) a member of its own species. For example, two human beings generate another human being and, although the generated human being is not identical to either of her generators in number, she is identical to both of them in species. Even a non-living substance such as some fire is generated from a fire that is identical to it in species, e.g. some fire catches some papers on fire, or a fire is divided into two fires.[146] But artefacts are not generated by, nor do they generate, objects that are identical to them in species and different from them in number. Therefore, artefacts are not substances.

Aquinas on the nature of artefacts

If artefacts are not substances for Aquinas, then what are they? One possible answer is that they are nothing but mental fictions. If this were true, then artefacts might be 'beings of reason' *(entis rationibus)*, but they would not enjoy extra-mental existence as trees and tigers do. In contrast to some contemporary authors who take a position like this,[147] Aquinas seems to say that artefacts do exist extra-mentally, in some sense. In what follows, I mention some of the different ways that Aquinas speaks about the non-substantial reality of artefacts.

One way that Aquinas talks about artefacts is as composites of form and matter. However, instead of being composed of substantial form and prime matter as substances are, artefacts are composites of a material substance (or an aggregate of material substances)[148] and an accidental form (or a set of accidental forms).[149] For example, take an artefact such as a bronze statue of George Washington. Here, the matter of the statue is a material substance, i.e. this piece of bronze, and the form of the statue is an instance of a certain sort of shape, perhaps *being in the shape of a man that resembles George Washington*, which is an accident (in the genus of quality) for Aquinas.[150] Aquinas also distinguishes between what might be called 'simple' and 'complex' artefacts: a simple artefact has one substance for its matter, e.g. some bronze, and an accident for its form of the whole, while a complex artefact has many substances as its matter and an accident (or perhaps even a set of accidents) as its form, e.g. as an axe is composed of some wood and some metal.[151]

Since artefacts are composites of a substance and an accidental form, they belong to a species of a genus of entities that Aquinas sometimes refers to as 'accidental beings'.[152] Not every accidental being is an artefact. For example, seated-Socrates is an accidental being, but not an artefact. Rather, an artefact is *a certain sort* of accidental being, presumably one that involves a human art. For instance, in the case of a bronze statue, it is only when a piece of bronze takes on a certain kind of shape, e.g. a shape such that the bronze signifies something beyond itself, in certain circumstances, e.g. where the efficient cause of the statue is a craftsperson – that we refer to that piece of bronze as a 'statue'.

Despite the difference between seated-Socrates (an accidental being that is not an artefact) and a statue, it will be useful to compare them for my purposes. What exists properly speaking in the case of seated-Socrates is Socrates, and 'seated-Socrates' refers to Socrates, albeit at a particular time – when he happens to have the accidental form, *being in a seated position*. Whether or not Socrates has this extraneous accident has no effect upon Socrates's numerical identity. Socrates does not go out of existence just in virtue of getting up from the seated position. However, the accidental being, seated-Socrates, does go out of existence when Socrates stands up. Likewise, for Aquinas a piece of bronze does not go out of existence just in virtue of its becoming statue-shaped. What exists properly speaking in the case of a statue is the piece of bronze that constitutes it; 'statue' refers to the accidental being that is a composition of a substance (the piece of bronze) and a certain accidental form, e.g. *being in the shape of a man*.[153]

In this chapter I have explained Aquinas's views on the different ways that material objects are composed. I've noted that material substances are unified material objects of the highest order, and so the integral parts of a substance x are not themselves substances, for they have their being and nature only in virtue of the substantial form of x. All material substances are composites of substantial form and prime matter, a substance's substantial form causing matter to be and to belong to a certain species, prime matter being the recipient of a substantial form and explaining the facts that a material substance's origin and destruction are natural occurrences and not miraculous ones.

Aquinas thinks there are many different kinds of material substances in the world. Some of these have functional integral parts such as hands and feet, whereas others have only non-functional (merely quantitative) integral parts. Some material substances are homeomerous (the elements), and so are not composed of parts belonging to kinds that differ from those substances in kind. Other material substances are heterogeneous (higher living organisms) and so are composed of a variety of different kinds of integral parts, each having its own distinctive function. Some material substances (compounds) are neither homeomerous – they are composed of elemental parts – nor are they heterogeneous since every integral part of these substances belongs to the same kind as the whole. There are some material substances – human beings – that can exist without having matter as a part. Although the metaphysical parts of most material substances are not themselves beings but rather those parts of a substance that account for the being and nature of those material substances, the substantial forms of human beings are subsistent, that is they are beings themselves and so can survive apart from the matter which they naturally inform.

Finally, Aquinas argues that artefacts are not substances. This is because the substances out of which artefacts are composed are not corrupted in virtue of the mere introduction of an artificial form. Although artefacts are real beings, they are, properly speaking, merely accidental beings: composites of a substance (or aggregate of substances) and an accidental form.

Having spoken about composition and parthood in this chapter, I turn in the next chapter to addressing another part of Aquinas's metaphysic of material objects that is important for offering a Thomistic response to the puzzles about material objects that raise the PMC: Aquinas's views on the identity and individuation of material objects through time and change.

Notes

1. What follows is not a complete account of Aquinas's theory of composition, a topic that deserves more attention than I can give here. I treat Aquinas's approach to composition only insofar as it helps set up my discussion of those modes of composition that are relevant for discussing the classical and contemporary puzzles about material objects that raise the PMC.

    Unless I say otherwise, I use 'composition' in the broadest possible sense in the sequel; that is, a composition is a whole of some sort made up of the relevant sorts of parts. Aquinas himself sometimes uses the term '*compositio*' in this broad sense, while sometimes he uses it in a restricted sense that signifies a particular *type* of composition.

2. See, e.g., ST Ia. q. 3, a. 7, c.

3. For example, at In Met. V, lec. 21, nn. 1093–7 Aquinas offers an extremely wide-ranging treatment of the different kinds of part (for additional discussions of the different senses of 'part', see also: SCG II ch. 72 (4), QDA a. 10, c., QDSC q. un., a. 4, c. and In Phys. I, lec. 3, n. 22). Aquinas explains the distinction between being one *per se* and being one *per accidens* at In Met. V, lec. 7. At In Met. VII, lec. 17, n. 1672 Aquinas discusses the distinction between composed objects that are one *simpliciter* vs composed objects that are one *secundum quid* (for additional treatments see also: SCG II ch. 56; ST IIIa. q. 2, a. 1, c. and ST IIIa. q. 73, a. 2, c).

4. SCG II ch. 56.

5. SCG IV ch. 35 (6).

6. ST IIIa. q. 2, a. 1, c.

7. Cf. ST IIIa. q. 90, a. 3.

8. Cf. Van Inwagen 1990a, pp. 18–20. Although Van Inwagen does not admit that the parts of material objects come in different kinds that are related analogously here, he does mention that the word 'part' is used analogously insofar as it is applied to material and non-material realities (e.g. as a stanza is a part of a poem). If I am right here about Aquinas's views, they can be seen as an extension of Van Inwagen's suggestion: just as 'part' is used analogously when speaking of the composition of material and non-material objects, 'part' is used analogously when speaking about the many different kinds of *material* parts.

9. 'Matter and form are said to be intrinsic to a thing, because they are parts constituting a thing' (DPN ch. 3); '[Matter and form] are related to a composed thing as parts to a whole, as the simple to the composed' (DPN ch. 4). See also SCG II ch. 54; QDA q. un., a. 1, ad13 and In VII Met., lec. 21, n. 1095.

10. I take the expression from Eleonore Stump. See, e.g., 2003, p. 35.

11. ST Ia. q. 76, a. 6, ad1: 'Una enim et eadem forma est per essentiam, per quam homo est ens actu, et per quam est corpus, et per quam est vivum, et per quam est animal, et per quam est homo.' See also SCG IV, chs 35 (7) and 81 (7).

12. See, e.g., ST Ia. q. 77, a. 6, c.

13. ST Ia. q. 76, a. 8, c.: 'Substantialis autem forma non solum est perfectio totius, sed cuiuslibet partis. Cum enim totum consistat ex partibus, forma totius quae est compositio et ordo, sicut forma domus: et talis forma accidentalis. Anima vero forma substantialis: unde oportet quod sit forma et actus non solum totius, sed cuiuslibet partis.'

14. See e.g.: DPN ch. 1 (3); DEE ch. 2 (20); DEE ch. 7 (101–2); SCG II ch. 50 (4); SCG II ch. 51 (2); SCG II ch. 57 (15); QDP q. 3, a. 8, c.; ST Ia. q. 76, a. 1, ob. 5; QDUVT q. un., a. 11, (near beginning of) c.; ST IIIa. q. 2, a. 1, c.; QDSC a. 3, ad12, In Met. VII, lec. 7, n. 1423, and In Met. IX, lec. 11, n. 1903.

15. See, e.g., SCG II ch. 54 (3).

16. Aquinas thinks that 'cause' and 'principle' can mean the same thing, but he also thinks that, properly speaking, there is a distinction to be made between these terms. However, the distinction is not relevant for my purposes here. For Aquinas's views on the relation between 'cause' and 'principle', see, e.g., DPN ch. 3 (18).

17. See, e.g.: DPN ch. 3 (15 and 17); In Sent. I, d. 23, q. 1, a. 1, sol. and SCG II ch. 68 (3).

18. See, e.g., DPN ch. 3 (18).

19. DEE ch. 7 (102). See also the texts in note 17 above.

20. SCG II ch. 71 (2): 'Forma autem unitur materiae absque omni medio: per se enim competit formae quod sit actus talis corporis, et non per aliquid aliud. Unde nec est aliquid unum faciens ex materia et forma nisi agens, quod potentiam reducit ad actum, ut probat Aristoteles, in VIII *Metaphysicae*: nam materia at forma habent se ut potentia et actus.' See also: SCG II ch. 58 (8); SCG II ch. 68 (3); ST Ia. q. 76, a. 7, c. and In Met. VIII, lec. 5, n. 1759.

21. See, e.g., QDP q. 3, a. 8, c.

22. See my general discussion of form in Chapter 3.
23. Bobik 1965, p. 74.
24. As we shall see in the sequel, matter too contributes to the being of a substance. But whatever matter contributes to the being of a material substance, it does not make that material substance to be actually what it is.
25. See also SCG II ch. 64 (2) where Aquinas argues that a harmony of parts is subject to degrees of more or less, whereas substantial forms are not – a substantial form is either present in matter or it is not.
26. In Met. VII, lec. 17, n. 1674: 'quod ipsum compositum non sit ea ex quibus componitur.'
27. See, e.g., Veatch, H. B. (1974) 'Essentialism and the problem of individuation' *Proceedings of the American Catholic Philosophical Association* 48: 64–73.
28. See, e.g., DPN ch. 6 (35); SCG II ch. 49 (4) and SCG II ch. 50 (3).
29. The *locus classicus* for this view in Aquinas is DEE ch. 3.
30. ST Ia. q. 76, a.1, c.: 'Sed considerandum est quod, quanto forma est nobilior, tanto magis dominatur materiae corporali, et minus ei immergitur, et magis sua operatione vel virtute excedit eam. Unde videmus quod forma mixti corporis habet aliquam operationem quae non causatur ex qualitatibus elementaribus.' See also SCG II ch. 68 (7–12); ST Ia. q. 78, a. 1, c. and QDSC q. un., a. 2, c.
31. Aquinas's understanding of the relative perfection of substantial forms shares one key feature with John Searle's understanding of emergent properties: the features of a substance, considered as a whole, are not reducible to the features of the integral parts of a substance, insofar as those parts are considered individually, or as a mere sum. See Searle, J. (1992) *The Rediscovery of the Mind*, Cambridge, MA: The MIT Press, pp. 111–12. For other contemporary discussions of emergent properties, see especially: O'Connor, T. (1994) 'Emergent properties' *American Philosophical Quarterly* 21: 91–104, and Hasker, W. (1999) *The Emergent Self*, Ithaca, NY: Cornell University Press, pp. 170–8. For the idea that Aquinas's view on the relative perfection of substantial forms is amenable to explication in terms of the contemporary notion of an emergent property, I am indebted to Eleonore Stump (see, for example: 2003, pp. 196–7).
32. SCG II ch. 68 (8). See also, ST Ia. q. 76, a. 1, c.
33. SCG II ch. 68 (10). See also, ST Ia. q. 76, a. 1., c.
34. SCG II ch. 68 (11). See also, ST Ia. q. 76, a. 1, c.
35. SCG II ch. 68 (12). See also, ST Ia. q. 76, a. 1, c.
36. See, e.g.: QDA q. un., a. 1; ST Ia. q. 75, a.1 and ST Ia. q. 76, a.1.
37. See, e.g., ST Ia. q. 75, a. 3.
38. Bazan, B. (1997), 'The human soul: form and substance? Thomas Aquinas's critique of eclectic Aristotelianism', *Archives d'Histoire Doctrinale et Littéraire du Moyen Age* 64: 95–126. For a text in Aquinas that suggests that this is an apt expression, see, e.g., ST Ia. q. 75, a. 2.
39. See, e.g., ST Ia. q. 75, a. 2, ad 1.
40. See, e.g., QDA q. un., a. 18, ad5.
41. See, e.g., DPN ch. 2 (12); QDP q. 3, a. 8; QDSC a. 3, ad12, and In Met. VII, lec. 7, n. 1423.

42. See, e.g., DPN ch. 2 (12); QDP q. 3, a. 8; QDSC a. 3, ad12, and In Met. VII, lec. 7, n. 1423.
43. See, e.g., ST Ia. q. 65, a. 4, c.
44. See, e.g., SCG II ch. 86 (6); QDP q. 3, a. 8, c., and In Met. VII, lec. 7, nn. 1423 and 1431.
45. See, e.g.: QDA 1; QDA 14 and ST Ia, q. 75, a. 2, c.
46. See, e.g., ST Ia. q. 89, a. 1, ad3.
47. See, e.g., SCG II ch. 86.
48. See, for example, SCG II ch. 87.
49. See, e.g., SCG II ch. 68 (3): 'Non autem impeditur substantia intellectualis, per hoc quod est subsistens, ut probatum est, esse formale principium essendi materiae, quasi esse suum communicans materiae.'
50. See, e.g., ST Ia. q. 75, a. 4, ad2.
51. See, e.g., SCG II ch. 57 and ST Ia. q. 75, a. 4, c.
52. See, e.g., ST IIa IIae. q. 83, a. 11. This view of Aquinas's will receive further support in my discussion of the identity of substances through change in Chapter 5.
53. This whole section is indebted to lectures I heard given by Eleonore Stump at Saint Louis University in February 2002. See also Stump 2003, pp. 52–4. For more on Aquinas's view that composition is not identity, see Chapter 6.
54. Stump 2003, p. 53.
55. For this view among contemporary philosophers, see for example, Van Inwagen 1990a, pp. 170–4 and Olson, E. (1997) *The Human Animal*, Oxford: Oxford University Press.
56. See, e.g., SCG II chs 68 and 83. See Bazan 1997 for more discussion.
57. See, e.g., SCG IV ch. 79 (10–12).
58. See, e.g., DPN ch. 1 (2).
59. See, e.g., DPN ch. 1 (2) and In Phys. I, lec. 13, n. 118.
60. DPN chs 1 and 2.
61. For further discussion of forms that do not configure anything – but are through themselves configurations – see Stump 2003, pp. 197–200.
62. See, e.g., DPN ch. 1 (6); DPN ch. 2 (10 and 11); In Phys. I, lec. 13, n. 118; In Phys. I, lec. 15, n. 139 and In Met., VIII, lec. 1, nn. 1688–9.
63. I use the example of the death of a living organism because it offers the most vivid example of what Aquinas calls 'the corruption of a substance'. Nonetheless, the points I make in what follows also apply in analogous fashion to the corruption of a non-living substance.
64. See, for example, Baker 2000, pp. 226–7.
65. See my discussion of these Thomistic positions in the sequel.
66. Cf. Aquinas's argument here for the existence of prime matter from the fact that material objects are generated and corrupted to Baker's argument that material objects have essential properties from the fact that material objects clearly have persistence conditions (2000, pp. 36–7).
67. For more discussion of this distinction, see: McGovern, M. (1987) 'Prime matter in Aquinas', *Proceedings of the American Catholic Philosophical Association* 61: 221–34.
68. See, e.g., DPN ch. 2 (11).

69. Zimmerman 1995, p. 76. See also my discussion of Dean Zimmerman's articles on material constitution in Chapter 2.

70. See, e.g., Aquinas's argument in DPN ch. 2 (12) that matter and form cannot themselves be things that are generated and corrupted on account of the fact that that would entail an infinite series of causes, and therefore explanations, making radical sorts of changes something unintelligible.

71. See, e.g.: ST Ia. q. 66, a. 2, c. and ST Ia. q. 76, a. 6 (end of c.).

72. DPN ch. 1 (2).

73. DPN ch. 1 (1 and 2).

74. DPN ch. 1 (2): 'Proprie loquendo, quod est in potentia ad esse substantiale, dicitur materia prima; quod vero est in potentia ad esse accidentale, dicitur subiectum. Unde dicitur quod accidentia sunt in subiecto; non autem dicitur quod forma substantialis sit in subiecto. Et secundum hoc differt materia a subiecto, quia subiectum est quod non habet esse ex eo quod advenit, sed per se habet esse completum; sicut homo non habet esse ab albedine. Sed materia habet esse ex eo quod sibi advenit, quia de se habet incompletum.'

75. DPN ch. 1 (2). See also: DPN ch. 1 (3); DPN ch. 2 (14); In Phys. I, lec. 13, n. 118; QDA q. 18 ad5; ST Ia. q. 50, a. 5, c.; QDSC a. 1 (2nd to last para. of the) c.; In Met. VII, lec. 2, n. 1292 and In Met. IX, lec. 9, n. 2289.

76. My interpretation of this passage was greatly aided by Bobik 1998, p. 5.

77. See, e.g.: DPN ch. 1 (2); ch. 2 (11); ch. 2 (14); In Sent. I, d. 39, q. 2, a. 2, ad4; In Sent. II, d. 12, q. 1, a. 4; In Sent. II, d. 34, q. 1, a. 4; QDV q. 8, a. 6; QDV q. 21, a. 2; SCG I ch. 17; SCG I, ch. 34; SCG II ch. 16; QDP q. 1, a. 1, ad7; QDP q. 3, a. 2; QDP q. 3, a. 5; QDA a. 12, ad12; ST Ia. q. 5, a. 3, ad3; ST Ia. q. 7, a. 2, ad3; ST Ia. q. 44, a. 2 ad 3; ST Ia. q. 48, a. 3; ST Ia. q. 66, a. 1 c.; ST Ia. q. 76, a. 6, c.; ST Ia. q. 115, a. 1, ad 2; QDSC q. un., a. 1, c.; QDSC q. un., a. 1, ad5 and In Met. VII, lec. 2, nn. 1285–6.

78. It may be that Aquinas admitted a second sort of substantial form at the time of writing his commentary on the *Sentences*. But, if he did, it is clear that he later rejects such a possibility unequivocally (since it is incompatible with his basic metaphysics). For discussion, see Wippel 2000, pp. 347 ff.

79. Baker 2000, p. 208.

80. See the section on the integral parts of a substance for detailed discussion of this idea.

81. I can't go into all the different arguments that Aquinas offers for the unicity doctrine. What follows are those that focus in particular on the nature of substantial form and matter. For more discussion of Aquinas's doctrine of the unicity of substantial form, as well as his reasons for holding it, see: Wippel 2000, pp. 295–375; Pegis, A. (1983) *St Thomas and the Problem of the Soul in the Thirteenth Century*, Toronto: Pontifical Institute of Medieval Studies, and Callus, D. (1961) 'The origins of the problem of the unity of form', *The Thomist* 24: 257–85.

82. ST Ia. q. 76, a. 4, c.: 'forma substantialis in hoc a forma accidentali differt quia forma accidentalis non dat esse simpliciter, sed esse tale: sicut calor facit suum subiectum non simpliciter esse, sed esse calidum. Et ideo cum advenit forma accidentalis, non dicitur aliquid fieri vel generari simpliciter, sed fieri tale aut aliquo

modo se habere: et similiter cum recedit forma accidentalis, non dicitur aliquid corrumpi simpliciter, sed secundum quid. Forma autem substantialis dat esse simpliciter: et ideo per eius adventum dicitur aliquid simpliciter generari, et per eius recessum simpliciter corrumpi. . . . Si igitur ita esset, quod praeter animam intellectivam praeexisteret quaecumque alia forma substantialis in materia, per quam subiectum animae esset ens actu; sequeretur quod anima non daret esse simpliciter; et per consequens quod non esset forma substantialis; et quod per adventum animae non esset generatio simpliciter, neque per eius abscessum corruptio simpliciter, sed solum secundum quid. Quae sunt manifeste falsa.' See also QDSC a. 1, ad9.

83. SCG II ch. 58 (5): 'Ab eodem aliquid habet esse et unitatem: unum enim consequitur ad ens. Cum igitur a forma unaquaeque res habeat esse, a forma etiam habebit unitatem. Si igitur ponantur in homine plures animae sicut diversae formae, homo non erit unum ens, sed plura. Nec ad unitatem hominis ordo formarum sufficient. Quia esse unum secundum ordinem non est esse unum simpliciter: cum unitas ordinis sit minima unitatum.' See also: ST Ia. q. 76, a. 3, c.; ST Ia. q. 76, a. 7, c., and QDSC a. 3, c.

84. As Wippel points out, these not only included Franciscan philosopher theologians such as John Pecham, Richard of Middleton, William of Ware, John Duns Scotus and William of Ockham, but also Henry of Ghent and Aquinas's teacher, Albert the Great (2000, pp. 312–13).

85. See, e.g., QQ 9 q. 2, a. 1, c. and ST Ia. q. 75, a. 2, ad1.

86. See, e.g.: In Met. V, lec. 21, nn. 1093–1097; ST IIaIIae. q. 48, a. 1, c., and ST IIIa. q. 90, a. 3.

87. ST IIIa. q. 90, a. 3. Thus, by calling a part p an 'integral' part of an object x, Aquinas does not mean to suggest that p is an essential part of x.

88. In Met. V, lec. 21, nn. 1093–7. Here Aquinas contrasts integral parts and wholes with universal parts and wholes. A universal whole, e.g. a logical genus (considered as a whole) can be predicated of its different species, i.e. the parts of the genus, as in the phrase 'human beings and horses are animals'.

89. Cf. In Met., lec. 21, nn. 1105–8. Here Aquinas makes a three-way division of the integral parts of a substance: (1) integral parts that compose a whole such that the position of those parts is not a proper accident of that whole; rather, the position of that whole's integral parts is completely a matter of that whole's relations to things extrinsic to itself, e.g. the water in a glass having its shape in virtue of the glass in which it is contained; (2) integral parts that compose a whole such that the (natural) arrangement of those parts is a proper accident of that whole, e.g. heterogeneous substances such as human beings have normal shapes, and (3) integral parts that compose a whole such that the position of those parts is not a proper accident of that whole, although the position of that whole's integral parts is not entirely due to that whole's relations to things other than itself, e.g. compound substances such as copper or clay that can hold their shape but do not have anything like a 'normal' shape. The distinction between (1) and (2) will remain largely irrelevant for my purposes here. However, I will speak about the differences between the integral parts of homeomerous and heterogeneous substances in the sequel.

90. See, e.g., In Met. VII, lec. 16, n. 1632.

91. See, e.g., QDSC q. un., a. 4, c.

92. See, e.g.: SCG IV ch. 49; CT ch. 211; QDUVI q. un., a. 2, c.; DQA q. 1, a. 1, c., ST Ia. q. 75, a. 2, ad1; ST Ia. q. 75, a. 4, ad2 and ST IIIa q. 2, a. 2, ad3.

93. SCG IV ch. 49 (13): 'Sed neque partes alicuius substantiae sic dicuntur particulares substantiae quasi sint per se subsistentes, sed subsistunt in toto.'

94. QDUVI q. un., a. 2, c. Cf. QQ IX, q. 2, a. 1, c. and QQ IX, q. 2, a. 2, c.

95. ST Ia. q. 75, a. 2, ad1. See also QDA q. 1, a. 1, c.

96. DQA q. 1, a. 1, c. See also: CT ch. 211; QDUVI q. un., a. 2, c. and ST Ia. q. 75, a. 2, ad1.

97. See e.g.: QDUVI q. un. a. 2, c.; QDUVI a. 2, ad 3; DQA a. 1, c.; QDA a. 1, ad 3; QDA a. 1, ad4; DQA a. 1, ad9; ST Ia. q. 75, a. 2, ad1; ST Ia. q. 75, a. 4, ad2 and ST IIIa q. 2, a. 2 ad3.

98. See, e.g., ST Ia. q. 76, a. 5, ad3.

99. See esp. QDA q. un., a. 1, ad4: 'although the human soul is able to subsist *per se*, it nevertheless does not have a complete species *per se*' (*licet anima humana per se posit subsistere, non tamen per se habet speciem completam*). Although Aquinas is here speaking explicitly of the human soul, as his frequent comparisons between integral parts and the soul suggest, the point made here that the human soul subsists *per se* but does not have a complete species *per se* applies equally to integral parts. See also: ST IIIa. q. 5, a. 3, c. and ST IIIa. q. 5, a. 4, c.

100. See, e.g.: In DA II, lec. 2, n. 239; SCG II ch. 57 (10); SCG II ch. 72 (3); ST Ia. q. 76, a. 8, c.; QDSC q. un., a. 4, c. and In Met. VII, lec. 11, n. 1519. As James F. Anderson suggests, Aquinas appears to have a passage from *De partibus animalis* (640b30–641a21) in mind when he makes this point (*Summa contra gentiles*. Book Two: *Creation* by Thomas Aquinas, Notre Dame, IN: University of Notre Dame Press, p. 213).

101. QDSC q. un., a. 2, ad5: 'nulla pars habet perfectionem naturae separata a toto.'

102. What *is* a severed limb for Aquinas? The answer seems to be that it is either a non-living compound of some sort, or else an aggregate of non-living substances. However, if Aquinas knew about modern cell biology, perhaps he would say that a severed limb is an aggregate of living things, namely an aggregate of cells.

103. ST Ia. q. 76, a. 8, c. See also: SCG II ch. 72 (3); SCG IV ch. 36 (7) and In Met. VII, lec. 13, n. 1588.

104. See, e.g.: CT ch. 212; In Met. V, lec. 21, n. 1102; In Met. VII, lec. 13, nn. 1588–91 and In Met. VII, lec. 16, nn. 1631–3.

105. In Met. VII, lec. 16, n. 1632.

106. See, e.g.: SCG II ch. 49 (3); SCG II ch. 56 (14); SCG II ch. 65 (3) and In Met. VII, lec. 13, nn. 1588–91.

107. In Met. VII, lec. 13, 1588: 'Duo enim, quae sunt in actu, numquam sunt unum actum; sed duo, quae sunt in potentia, sunt unum actum, sicut patet in partibus continui.'

108. In Met. VII, lec. 13, 1588: 'Unumquodque enim dividitur ab altero per propriam formam. Unde ad hoc quod aliqua fiant unum actum, oportet quod omnia concludantur sub una forma, et quod non habeant singula singulas

formas, per quas sint actua. Quare patet, quod si substantia particularis est una, non erit ex substantiis in ea existentibus actu[.]'

109. In Met. VII, lec. 13, n. 1589.

110. For the sake of convenience I talk in this section as though Aquinas is right about the nature of elements and compounds.

111. See esp. In Met. V, lec. 21, n. 1105–8 and In Met. VII, lec. 16, n. 1634. Cf. In Phys. I, lec. 9, n. 65. See also note 89 above.

112. In Met. VII, lec. 11, n. 1519.

113. In Met. VII, lec. 16, n. 1634: 'Quamvis enim omnes partes sint in potentia, tamen maxime poterit aliquis opinari partes animatorum et partes animae esse propinquas, ut fiant actu et potentia, idest ut sint in potentia propinqua actui. Et hoc ideo, quia corpora animata sunt corpora organica habentia partes distinctas secundum formam.' Cf. In Phys. I, lec. 9, n. 65.

114. Van Inwagen 1981. Van Inwagen thinks that even 'functional integral parts' raise the PMC. This is why, as we have seen in Chapter 2, he thinks that the only existing integral parts of complex material objects are cells and physical simples.

115. See, e.g.: In Met. V, lec. 21, nn. 1105–7 and In Met. VII, lec. 16, n. 1633. See also my discussion of the individuation of non-living substances in Brown, C. (2001) 'Aquinas on the individuation of non-living substances', *Proceedings of the American Catholic Philosophical Association* 75: 237–54.

116. See, e.g.: SCG II ch. 72 (4); SCG II ch. 86 (3); QDP q. 3, a. 12, ad5; ST Ia. q. 76, a. 3, c.; In DA I, lec. 14, n. 208, and In Met. VII, lec. 16, n. 1635. Aquinas also mentions cases where wholes become parts, that is where what is actually existent as a whole becomes potentially existent as a functional part (see In Met. VII, lec. 16, n.1636). See my discussion of such possibilities in Brown 2001.

117. Aquinas's understanding of integral parts also has implications for his views on the individuation of non-living substances. I don't talk about these here. See Brown 2001 for a detailed discussion.

118. The expression is Peter van Inwagen's. See Van Inwagen 1990a.

119. See, e.g.: In Sent. II, d. 3, q. 1, a. 4; In BDT q. 4, a. 3, c.; SCG II ch. 65 (3); QQ 1 q. 4, a. 1, ad3; ST Ia. q. 52, a. 3, ob. 1; ST Ia. q. 75, a. 4, ad4; DME (5) and ST IIIa. q. 75, a. 1, ob. 3.

120. However, as we'll see, Aquinas also denies that an aggregate of substances is an individual thing in the same sense as a material substance is an individual thing.

121. In Met. V, lec. 4, n. 800. However, Aquinas does think that certain quantities of a compound kind are such that they cannot hold the form of the compound in question and will therefore be dissolved immediately into instances of the elemental kinds that compose them. Thus, although every integral part of a non-living compound substance x belonging to kind K is some K – at least while it is a part of x – some parts of x are such that they will be dissolved into instances of the elemental kinds that compose K-compounds upon being separated from x. See my discussion of the extension of 'material substance' in Chapter 3 for additional discussion and texts.

122. I can't examine here all of the approaches to elemental presence that Aquinas considers throughout his career. For Aquinas's evaluation of a position on

elemental presence that he attributes to Avicenna, see: In II Sent. d. 12, q. 1, a. 4; BDT q. 4, a. 3, ad6; QQ 1 q. 4, a. 1, ad3; QDA q. un., a. 9, ad 10; ST Ia. q. 76, a. 4, ad4 and DME (2–5). For Aquinas's evaluation of a position on elemental presence that he attributes to Averroes, see: QQ 1 q. 4, a. 1, ad3; ST Ia. q. 76, a. 4, ad4; QDA q. un., a. 9 and DME (7–14).

123. See, e.g., BDT q. 4, a. 3, ad6; QQ 1 q. 4, a. 1, ad3; ST Ia. q. 76, a. 4, ad4 and DME (5–6).

124. Bk I, ch. 10, 327b29–31.

125. See e.g.: QQ 1 q. 4, a. 1, ad3; SCG II ch. 56 (4); SCG IV ch. 35 (8); QDA q. un., a. 9, ad 10; ST Ia. q. 76, a. 4, ad4; ST IIIa. q. 2, a. 1, c. and DME (18). For some discussion of the development of Aquinas's thought on elemental presence, see: Baldner, S. (1999) 'St Albert the Great and St Thomas Aquinas on the presence of elements in compounds', *Sapientia* 54: 41–57.

126. DME (16–18).

127. See my discussion of proper accidents in Chapter 3.

128. In Met. VII, lec. 16, n. 1633.

129. See, e.g., ST Ia. q. 75, a. 1, ad2.

130. I was greatly aided in making these connections by Bobik 1998, p. 125.

131. Decaen, C. (2000) 'Elemental virtual presence in St Thomas', *The Thomist* 64: 271–300, p. 298.

132. In Met. VII, lec. 16, n. 1633. See also: SCG IV ch. 35 (8); ST IIaIIae q. 48, a. 1, c. and ST IIIa. q. 2, a. 1, c.

133. Although, recall that elements can have *elements* as *integral* parts. For example, a puddle of water (call it 'x') that counts as an instance of an element could be divided into two puddles (call these 'y' and 'z'). y and z are substances in potency while they are parts of x, but they are actual substances after they are divided from and are no longer parts of x. Nonetheless, when y and z are parts of x, they are not elemental parts of x. This is because elemental parts compose something that differs from those parts in kind (which is not true in this case) and elemental parts are present in the whole of the substance of which they are parts (also not true in this case).

134. See, e.g., In BDT q. 5, a. 3, c.: 'per se competit homini quod inveniatur in eo anima rationalis, et corpus compositum ex quatuor elementis; unde sine his partibus homo intelligi non potest: et sic oportet poni in definitione eius.' See also CT ch. 170.

135. See also Aquinas's discussion of the glorified body at ST Suppl. qq. 79–85 and SCG IV chs 84–6.

136. See, e.g.: Baker 2000, pp. 22–5; Burke 1997, p. 11, and Hirsch, E. (1999) 'Identity in the Talmud', *Midwest-Studies-in-Philosophy* 23: 166–80.

137. See, e.g.: SCG IV ch. 35 (7); In Phys. I, lec. 12, n. 109; In Phys. II, lec. 1, nn. 142 and 145 and In Phys. II, lec. 2, nn. 149 and 154.

138. See, e.g.: DPN ch. 1 (6); In DA II, lec. 1, n. 218; In DA II, lec. 2, nn. 235–7; In Phys. I, lec, 2, n. 14; In Phys. I, lec. 12, n. 108; In Phys. II, lec. 2, n. 149; In Met. VII, lec. 17, n. 1680 and ST IIIa. q. 2, a. 1.

139. DPN ch. 1 (6): 'Sicut quando ex cupro fit idolum, cuprum quod est in potentia ad formam idoli, est materia; hoc autem quod est infiguratum sive

indispositum, est privatio; figura autem a qua dicitur idolum, est forma; non autem substantialis, quia cuprum ante adventum illius formae habet esse in actu, et eius esse non dependet ab illa figura, sed est forma accidentalis. Omnes enim formae artificiales sunt accidentales.' See also In Phys. II, lec. 2, n. 149 and In DA II, lec. 1, n. 218.

140. For example, this is part of Michael Burke's solution to the problem of material constitution where artefacts such as statues are concerned (1994a, pp. 595–6).

141. See, e.g.: SCG IV ch. 35 (7); In DA II, lec. 1, n. 218; In Phys. II, lec. 1, nn. 142 and 145; In Phys. II, lec. 2, nn. 149 and 154; In Met. VII, lec. 17, n. 1680 and ST IIIa. q. 2, a. 1.

142. In Phys. II, lec. 1, nn. 142 and 145.

143. See, e.g., ST Ia. q. 75, a. 2, ad2.

144. Lynne Rudder Baker would agree that the axe does not undergo a substantial change when the foreigners begin using the axe as a doorstop, but she would say this on the grounds that *once a thing is knighted an 'axe', that thing remains an axe for as long as it exists* (Baker 2000, p. 38). I don't see that she provides any reason for accepting this claim. Why should we privilege the thoughts of the original creators of the axe over the thoughts of those who later find the axe?

145. In Phys. II, lec. 2, nn. 149 and 154.

146. See, e.g.: ST Ia. q. 119, a. 1, ad5.

147. See, e.g., Van Inwagen 1990a and Hoffman and Rosencrantz 1997.

148. See, e.g.: DPN ch. 1 (6); In DA II, lec. 1, n. 218; In Phys. I, lec. 12, n. 109 and In Met. VII, lec. 3, n. 1710.

149. See, e.g.: DPN ch. 1(6); In Sent. IV, d. 44, q. 1, a. 1, sol. 2, ad4; SCG II ch. 72 (3); SCG IV c. 35; In DA II lec. 1, n. 218; lec. 2, nn. 235–7; ST Ia. q. 76, a. 8, c. and ST IIIa q. 2, a. 1.

150. See, e.g., DPN ch. 1.

151. See, e.g., In Met. V, lec. 3, n. 779.

152. For passages where Aquinas mentions accidental being, see, e.g.: DPN ch. 1; In I Sent., d. 8, q. 5, a. 2, c.; ST Ia. q. 9, a. 2, c.; In I Phys., lec. 2, n. 14; In I Phys., lec. 12, n. 108; In II Phys., lec. 1, n. 145; In V Met., lec. 9, nn. 885–7; In VI Meta., lec. 2, n. 1179 and In VI Met., lec., 4, nn.1241–4. The passages at In VI Met. in particular suggest that accidental beings are not beings in the proper sense. See also DEE ch. 7, where Aquinas speaks about the composition of a substance and an extraneous accident as something *unum per accidens*.

153. See, e.g., In DA II, lec. 2, n. 237.

5

Aquinas on Identity, Individuation and Material Composition

In this chapter I discuss Aquinas's views on the identity relation and, for him, the closely related issue of what explains why a material object is an individual thing distinct from all other individual objects. The chapter has three sections. In the first section I explain Aquinas's views on the identity of material substances through time and change. The second section explains Aquinas's views on the individuation of material substances, with particular attention paid to difficult cases for Aquinas's views. Finally, I close the chapter with a section that develops a Thomistic position on the identity and individuation of artefact objects.

Aquinas on the identity of material substances through time and change

As we have seen, Aquinas thinks that material substances are material objects in the most primary sense. It therefore makes sense to begin my discussion of Aquinas's views on the identity of material objects by focusing on the identity of material substances. In this section I first say a few things about the notion of 'relative identity' and argue that Aquinas is a relativist about identity only in a very innocuous sense. I then go on to show that Aquinas has the view that the substantial form of a material substance explains how that material substance can remain numerically identical through time and change.

Relative identity in contemporary philosophy

As I mentioned in Chapter 1, some contemporary philosophers solve the problem of material constitution by denying the following (logical) intuition:

(IMO6) Necessarily, identity is a transitive relation.

Those who deny IMO6 are spoken of as 'relativists' about the identity relation by contemporary philosophers. There are at least three different meanings given to the expression 'relative identity' by contemporary analytic philosophers.[1]

The most common meaning given to the locution 'relative identity' by contemporary philosophers picks out a view that we can label 'sortal relativism'. The advocate of sortal relativism accepts the following:

(SR) It is possible that, for some material objects x and y, and for some properties, F and G, objects x and y are the same F, but not the same G.

A second way of thinking about the identity relation as relative is to think that identity is relative to possible world (or time). Those who take identity to be relative to possible world think the following claim is true:

(PWR) It is possible that $x = y$ in possible world W, x is not identical with y in another world $W1$, and x and y both exist in world $W1$.

We can call the position that takes identity to be relative to possible worlds, 'possible-world relativism'.

In a manner similar to the possible-world relativist, some philosophers take the even more radical view that identity is relative to time. We can call the position that takes identity to be relative to time 'temporal relativism'. Temporal relativists accept the following:

(TR) It is possible that $x = y$ at time t in possible world W, x and y are not identical at $t + 1$ in W, and x and y both exist at $t + 1$ in W.

To these three senses of 'relative identity' I would like to propose a fourth. This fourth sense of 'relative identity' is related to adopting a certain interpretation of the Law of the Indiscernibility of Identicals (LII). Let's begin with how one contemporary philosopher has formulated LII:

(LII) For any x and y, x is identical to y only if x and y have all and only the same properties.[2]

The following question might occur to the reader of LII: what meaning is to be given to 'identical' in the left-hand side of LII? It seems clear that many philosophers take the meaning of 'identical' in LII to be *numerical* identity.[3] In that case, LII should be read as:

(LII*) For any x and y, x is numerically identical to y only if x and y have all and only the same properties.

But LII* seems to run into the problem that it makes change over time look impossible. For according to LII*, in order for x and y to be numerically identical, x and y must have all and only the same properties. But if an object x is to change, x must possess some property P after the change that it did not possess before that change. Therefore, LII* and the possibility of change over time appear to be logically incompatible.

Now contemporary philosophers have spilled quite a bit of ink trying to solve this particular problem, which just goes to show that many contemporary philosophers are indeed committed to LII*.[4] But consider an alternative interpretation of LII, one that I will call a 'relativist' interpretation. The philosopher who is a 'relativist' about identity in this sense thinks that terms such as 'identity' and 'identical' are ambiguous in the following way: '$x = y$' might mean 'x and y are *absolutely* identical, that is, x and y have all and only the same properties', or '$x = y$' might mean, 'x and y are *numerically* identical, that is, x and y are the same object, where x and y can be the same object without necessarily sharing all and only the same properties'. So the relativist about identity in the sense I am describing here makes a distinction between *absolute identity* (x and y being exactly the same) and *numerical identity* (x and y being merely the same object, where x and y might have different accidental properties). Although all objects that are absolutely identical are therefore numerically identical, it is not the case that x and y are absolutely identical just because they are numerically identical. For lack of a better term, we can call the view that takes identity claims to be ambiguous in this way 'LII relativism'. The LII relativist therefore disambiguates LII as follows:

(LII**) For any x and y, x is absolutely identical to y only if x and y have all and only the same properties.[5]

Armed with the distinction between absolute and numerical identity the LII relativist can make sense of our common-sense notion that objects remain the same despite undergoing certain accidental changes. Although my bike today has properties different from those that it had yesterday – my bike is not absolutely the same today as compared to yesterday – my bike yesterday and my bike today are nonetheless numerically identical.

Is Aquinas a relativist about identity?

Having shown that there are at least four different meanings that can be given to the locution 'relative identity', and since claiming that identity is relative (in at least some of these senses) will affect one's approach to solving the PMC, it is reasonable to ask at this point whether Aquinas is in any sense a relativist about identity.

Aquinas himself characterizes identity as a kind of unity or union.[6] For x and y to be identical or the same is for x and y to be *one* in a certain sense.[7] Specifically, Aquinas thinks that x and y are identical if and only if they are *one in substance*, that is, one, and only one, substance. Thus, for Aquinas, identity is equivalent to substantial unity. Properly speaking, only substances can be identical for Aquinas.[8]

To speak of identity in terms of unity or union as Aquinas does is help-ful since it highlights the fact that identity is a *relation*. Now every relation has at least two *relata*, and since a relation can have both extra-mentally existent things (particular substances) and merely logical beings (or universal substances) as its *relata*, this suggests there are a number of different ways that we might speak about the identity of substances *x* and *y*. Aquinas men-tions three ways: (1) *x* and *y* can be numerically the same substance (one in *suppositum*), although *x* and *y* are referred to by terms that differ in sense, e.g. Cicero and Tully; (2) *x* and *y* can be the same in species, while differing numerically, e.g. Cicero and Seneca, and; (3) *x* and *y* can be numerically the same, where *x* and *y* are referred to by different tokens of the same term type, e.g. in the statement 'Cicero is identical with Cicero'.[9]

Now among the different senses of relative identity that I have examined, it should be clear that Aquinas is an LII relativist. Aquinas thinks that it is obvious that Socrates can gain and lose properties without losing his numerical identity (and therefore his existence). Although baby Socrates and Socrates the philosopher are not absolutely identical — Socrates the philosopher has many features that baby Socrates does not have (and vice versa) — they are nonethe-less one and the same object, that is, they are numerically identical.

Although Aquinas is a relativist about identity in the sense that he distin-guishes between complete sameness with respect to a substance's properties over time and the numerical identity of a substance over time, this view of his does not cause him to deny any of the common-sense assumptions about mate-rial objects that are at issue in the puzzles having to do with material constitu-tion. Indeed, his views are commonsensical on this point. Thus, Aquinas's espousing of LII relativism does not affect his solution to the PMC in the sense that it entails he give up a common-sense intuition about material objects (or logic).

More to the point where the PMC is at issue is the question whether Aquinas is a sortal relativist, a possible-world relativist or a temporal relati-vist. Of course, Aquinas does not speak about necessity and contingency in terms

of possible worlds. In fact, it has been argued that the medieval notions of modality employed by Aquinas are very different from those used in con-temporary modal semantics.[10] Furthermore, it seems that possible-world relativism and temporal relativism are relatively new philosophical posi-tions. Since nothing Aquinas says about identity suggests an allegiance to a position analogous to possible-world relativism or temporal relativism, I think it is safe to conclude that he is not a relativist about identity in either of these senses.

There remains the more interesting question whether Aquinas is a sortal relativist. It might appear that he is, given some of the things I've said above about Aquinas's views on identity. Indeed, as we have seen, Aquinas thinks that 'identity' can have different meanings, so that the following statement is ambiguous:

(a) $x = y$.

Aquinas thinks that (a) is ambiguous since, although any identity statement about x and y entails that x and y are identical in one sense, they might not be identical in another sense, e.g. Cicero and Seneca are identical in species, but not in number; baby Socrates and Socrates the philosopher are numerically identical but not absolutely identical.

However, as Aquinas makes clear, all identity statements have substances of some sort as their *relata*. This, I think, is enough to see that Aquinas is not a sortal relativist. To see why, consider that, although the sortal relativist too thinks (a) is ambiguous, the sortal relativist thinks that (a) should be revised along the following lines:

(b) x is the same F as y.

Furthermore, according to the sortal relativist, (b) is compatible with

(c) x is not the same G as y.

But what kinds of things stand in for 'F' and 'G' for the sortal relativist? Clearly, specific substance-sortals such as 'human being' and 'piece of cellular tissue' are the sorts of things that stand in for 'F' and 'G'. However, if the sortals that stand in for 'F' and 'G' are substance sortals, Aquinas would not agree that (b) and (c) could both be true. If Socrates at time t is the same particular substance as Socrates at $t+1$, then it is not possible that Socrates at t and Socrates at $t+1$ are, for example, different animals or different bodily substances. For example, if Socrates is the same human being at two times, then Socrates will be the same animal at those two times. As Aquinas points out in one place, whatever is numerically identical, is also thereby specifically identical.[11] But let's take examples of sortals where both sortals pick out individual substances belonging to different species, e.g. the standard sortal relativist suggestion that Socrates at two times is the same human being at those two times but not the same piece of cellular tissue at those two times. The problem with this supposed example of sortal relativity from Aquinas's perspective (and any other relevantly similar example) is that one of these sortals is not really a substance sortal. For example, pieces of cellular tissue are not substances for Aquinas. At most, such things are parts of substances (if they exist at all). But identity statements make claims about how things are one (the same) in substance. This requires that the sortal in question be a substance sortal (which 'celluar tissue', at least in the context, is not). So for Aquinas, it wouldn't make sense at all to affirm or deny 'Socrates *is* the same piece of cel-lular tissue at two different times'. Thus, since Aquinas thinks that a conjunc-tion of statements having the forms of (b) and (c) is either going to be necessarily false or else ill-formed, and sortal relativism requires that a con-junction of statements having the forms of (b) and (c) is something possible, I conclude that Aquinas rejects sortal relativism.

As we have seen, the only way that Aquinas can be construed as a relativist about identity is with respect to LII, and LII relativism is an innocent sort of relativism where the PMC is concerned since it is not in conflict with common-sense intuitions about material objects or logic.

We have seen Aquinas say that compound material substances are able to preserve their numerical identity through time and change. What is the cause of material substances x and y being numerically identical for Aquinas? I turn to addressing this question in the next section.

Aquinas on the numerical identity of material substances

Aquinas clearly has the view that the identity of a substance x is rooted in the identity of x's metaphysical parts. Why a substance's metaphysical parts, and not its integral parts? Just as Aquinas accepts from our common way of under-standing the world that material substances have essential and accidental fea-tures, he also accepts that all material substances can survive certain losses, gains or replacements of their integral parts.[12] For example, a tree can survive losing its leaves every fall. But what about those integral parts that a material substance can't lose without thereby being corrupted? For example, a dog certainly requires its brain in order to remain in existence. However, to say that the substantial form of a material substance x explains x's remaining numerically identical through time and change does not mean that something can't happen to x such that it loses its substantial form, thereby causing x to go out of existence. Furthermore, as we saw in Chapter 4, all of a material sub-stance's integral parts have their being and species in virtue of the substantial form of that substance. So it is more correct to say that a dog's substantial form and not some integral part of the dog is what accounts for the identity of that dog through time and change. That the substantial form of a material sub-stance x is the cause of x's being and x's belonging to the species that it does – not to mention the being and species of x's integral parts – suggests that the substantial form of a material substance x is a necessary and sufficient condi-tion for preserving the identity of x for Aquinas. Indeed, according to one recent interpreter of Aquinas on identity, Aquinas accepts something like the following:

(AI) For any material substances x and y, x is numerically identical to y if and only if the substantial form of x is numerically identical to the substantial form of y.[13]

However, there are texts that suggest that Aquinas rather accepts the following slightly different claim:

(AI*) For any material substances x and y, x is numerically identical to y if and only if the substantial form of x is numerically identical to the substantial form of y and the matter of x is numerically identical to the matter of y.

For example, when Aquinas speaks about the identity of human beings in the context of treating the Christian doctrine of the resurrection of the dead, he appears to accept the soundness of the following argument:

(IR)(1) In order for Socrates at the resurrection to be numerically identical to the Socrates who once lived on earth, the essential principles of Socrates at the resurrection must be the same as the essential principles of the Socrates who once lived on earth. (2) But the essential principles of any human being are a substantial form and matter. (3) Therefore, for Socrates at the resurrection to be numerically identical to the Socrates who once lived on earth, the Socrates at the resurrection must have the same substantial form and matter that Socrates had on earth.[14]

Since premise 1 of IR appears to assume that a substance remains numerically the same if and only if it retains the same essential principles, and premise 2 makes it clear that these principles are a substance's metaphysical parts – both a substance's substantial form and matter – IR makes it appear as though Aquinas accepts AI* rather than AI.

So does Aquinas accept AI or AI*? Or are his views inconsistent on this score? I now want to offer a number of reasons for why clearly Aquinas accepts AI and not AI*. I'll also explain why IR is actually consistent with AI.

First of all, consider those material substances whose substantial forms are non-subsistent (the substantial forms of non-human material substances). AI provides necessary and sufficient conditions for the identity of non-human material substances for the following reason: if non-human material substances x and y have the same substantial form, then they automatically have the same matter, since (as we shall see in the sequel) substantial forms themselves are individuated by matter. The presence of the same substantial form is enough to guarantee that a non-human material substance retains its numerical identity through time and change.

However, as we have seen, Aquinas thinks that some substantial forms of material substances – human souls – can exist apart from matter. Aquinas's views on the ontological status of human beings after death and before the general resurrection provide further reason for thinking that AI is a more precise formulation of Aquinas's general views on the identity of material substances than AI*. For consider whether or not a consequence of AI* isn't that Socrates *goes out of existence* at the moment of his biological death. When the soul is separated from the body at biological death according to Aquinas, it continues to exist; the soul of a human person is not annihilated at death.[15] Thus, the human soul is literally 'separated' from the body at death according to Aquinas: it exists for a time in a separated state, that is the soul exists for a time without its material (bodily) complement. Although it is clear that *Socrates's soul* exists after Socrates dies for Aquinas, if Aquinas accepted AI*, then he would seem to be committed to the view that *Socrates* does not exist after his biological death (at least prior to the general resurrection). This is because all that exists after Socrates's death is a soul, and AI* implies that no *material being* can exist without matter (and therefore a body) as a component part. We can call this account of Aquinas's understanding of the ontological status of Socrates after his biological death the *corruptionist account*, since this account of Aquinas's views has it that Socrates is corrupted at death in the sense that at his biological death Socrates goes out of existence.

In addition to argument IR, the corruptionist account seems to receive support from Aquinas's view that Socrates is not identical to his soul.[16] Since Socrates does not have a body after his biological death (at least before the resurrection), and Socrates is not identical to his soul, Socrates would seem to go out of existence at his biological death. This latter argument for the corruptionist account also assumes that Aquinas accepts (something like) AI*.

According to another interpretation of the ontological status of a human being after death and before the resurrection – call it the *alternative account* – *Socrates* is *not* corrupted at death; rather, only Socrates's *body* is.[17] Since Socrates's soul survives Socrates's death, and because a human soul is sufficient to preserve the numerical identity of a human being through time and change, *Socrates* survives his biological death. Although Socrates is never *identical* to his soul – even in the ante-resurrection, post-death state – he can be *composed* of only his soul. Indeed, this is Socrates's situation in the ante-resurrection, post-death state. Notice that according to the alternative account Aquinas does not accept AI*.[18]

Aquinas clearly has the view that Socrates is not identical to the physical parts that compose him at any given time. This is because Socrates remains numerically the same human being – and therefore numerically the same *being* – through the loss and gain of integral parts such as *this carbon atom* or even *this arm*. For Aquinas, Socrates retains his numerical identity through the loss and gain of integral parts in virtue of his substantial form, that is, his soul.[19] Therefore, Aquinas clearly has the view that composition is not identity where Socrates's physical parts are concerned. What Aquinas's views on personal identity and the resurrection show – according to the alternative account proposed here – is that Aquinas thinks that composition is not identity even where the *metaphysical* parts of a human being are concerned (but only in cases involving human beings and their souls). Socrates is not identical to his soul but Socrates can be composed of just his soul.[20]

Now although the corruptionist account of Socrates's death is clearly compatible with the premises and conclusion of argument IR, so is the alternative account. First of all, notice that IR is an argument having to do with numerical identity *at the resurrection*. The argument does not show that a human being loses her identity if she loses her body. What the argument shows is that a human being will lose her identity if her soul is joined to matter at the resurrection that is numerically different from the matter she had in her earthly life.[21] But nothing about the alternative account of Socrates's death conflicts with this view.

A casual reading of premise 2 of IR offers another explanation for the immediate appeal of the corruptionist account. Since the soul of Socrates is not united to matter in the post-death, ante-resurrection state, and Socrates's soul constitutes only one of his essential principles, Socrates must be corrupted at death.

Indeed, Aquinas's claim that Socrates's soul is not a human being *should* be read as saying 'Socrates's soul alone is *not identical* to a human being'. The advocate of the alternative account accepts this reading, but thinks – because composition is not identity – that saying that Socrates is not identical to his soul is not incompatible with the view that *Socrates* may exist for a time without his body. Socrates's species is *homo*, and the definition Aquinas accepts of *homo* is *rational animal*.[22] So Socrates is – where 'is' is understood according to the 'is' of identity – a rational animal, and rational animals are *normally* composed of both form and matter. Like the corruptionist account, the alternative account accepts all of these views. But in contrast to the corruptionist account – which says that Socrates goes out of existence at death, Socrates's body no longer existing – the alternative account has it that, although Socrates's soul is not identical to Socrates, Socrates can be composed of only his soul, that is, Socrates's soul is sufficient to preserve Socrates's identity, even in the post-death, ante-resurrection state. Notice that the alternative account is not incompatible with premise 2 of IR: Socrates, because he is a human being, is a rational animal, and rational animals are such that they are (normally) composed of a soul and a body.

So far I have shown that, like the corruptionist account of Socrates's death, the alternative account is compatible with Aquinas's accepting that argument

IR is sound. Now I want to turn to developing an argument that the alternative account is a better interpretation of Aquinas's views than is the corruptionist account. This will further bolster my contention that AI and not AI* is a more accurate formulation of Aquinas's views on the identity of material substances through time and change.

There is a serious problem with the corruptionist account of Aquinas's views on the ontological status of Socrates after death and before the resurrection: it has the unwelcome consequence that Aquinas is committed to what I shall call the Intermittent Existence Thesis:

> (IE) It is possible that x exists at time t, x goes out of existence at $t + 1$, y begins existing at $t + 2$, and y is numerically identical to x.

Why think Aquinas is committed to IE if he holds to the corruptionist account? Because, according to the corruptionist account, Socrates goes out of existence at his biological death and the Christian doctrine of the resurrection of the dead (which Aquinas, of course, accepts) entails that Socrates at the resurrection is numerically the same human being as Socrates before his biological death. But then it is the case that Socrates exists for a time (before his biological death), goes out of existence at his biological death and *then* comes back into existence at the general resurrection. Therefore, the traditional Christian understanding of the resurrection of the dead and the corruptionist account of Socrates's death together entail IE.

I take Aquinas's supposed acceptance of IE by way of the corruptionist account to be an *unwelcome* consequence for the simple reason that Aquinas himself rejects IE, and thus ascribing the corruptionist account to Aquinas means that Aquinas's views on personal identity are inconsistent with his general metaphysics. Even if ascribing a plainly false view to Aquinas doesn't give his interpreter pause (after all, there may be some debate over what is plainly false, e.g. many contemporary philosophers accept IE), ascribing an inconsistent set of views to a thinker of the calibre of Aquinas certainly should. Of course, if there were no viable alternative to the corruptionist account – that is, no other way of understanding all the things Aquinas has to say about the nature of human beings, personal identity and the resurrection of the dead – that would be a different story. But I think I have shown that Aquinas need not be committed to the view of Socrates's death that is entailed by the corruptionist account – and with it the truth of IE – since there is another plausible Thomistic account of the ontological status of Socrates upon his death, namely, the *alternative account*.

One problem then with the corruptionist account is that it entails IE.[23] The alternative account does not. If Aquinas does reject IE, this would constitute a good reason for favouring the alternative over the corruptionist account. But does Aquinas reject IE?

In fact, Aquinas seems to do just that in a place where he is answering an objection to the possibility of the resurrection of the dead. The objection runs as follows:

> (1) That which is not continuous does not appear to be the same in number (this is clear in cases involving not only sizes and motions, but also in cases involving qualities and forms: for if, after healing, someone becomes sick and then is healed again, the health which returns will not be the same in number). (2) But it is clear that through death the being of a human being is taken away, since corruption is a change from being to non-being. (3) Therefore, it is impossible that the being of a human being be repeated the same in number. (4) Therefore, neither will a human being [return] the same in number [at the resurrection], since those things which are the same in number are the same according to their being.[24]

Aquinas answers this objection to the Christian understanding of the resurrection by rejecting premise 2 of the objection above: the substantial being of a human being is *not* taken away at death – the human soul is alone sufficient to preserve the identity and existence of a human *substance* even upon the corruption of the body at biological death. Here is Aquinas's reply to the objection:

> It is clear that the being of matter and form is one. For matter does not have actual being except through form. But the rational soul differs from other forms on this point. For other forms do not exist except in their union with matter: for they do not exceed matter either in being or operation. But it is clear that the rational soul exceeds matter in its operation. For it has a certain operation in which no organs of the body participate, namely understanding. Hence [the rational soul's] being also does not exist only in its union with matter. Therefore, [the rational soul's] being which was [the being] of the composite, remains in [the rational soul] upon the dissolution of the body. And upon the restoration of the body at the resurrection, [the being of the composite] returns to the same being that remained in the soul.[25]

If Aquinas accepted IE, he would have rejected premise 1 of the objection – that an object's being could be taken away and then at some later time be restored. He does not do so, I suggest, because he accepts what premise 1 implies: IE is false; intermittent existence is a metaphysical impossibility.[26] Since the corruptionist account entails IE and the alternative account does not, Aquinas's rejection of IE shows that the alternative account constitutes a more charitable interpretation of Aquinas's views on personal identity.[27]

The question of the possibility of IE aside, note again that Aquinas's answer to the objection above also argues that the separated soul after death retains the same *being* that the composite had prior to death. This is why the resurrected person can be numerically the same as some person prior to her death. But this is just what one would expect on the alternative account, although not on the corruptionist account. Clearly the *being* at issue in this passage is substantial and not accidental being. If the separated soul retains the same substantial being that was enjoyed by the composite before death, then Socrates does not undergo substantial change at his death. But Aquinas says in the passage above that Socrates, for example, does in fact retain his substantial being at death – in virtue of his soul's incorruptibility – despite the fact that his soul is separated from matter at his death. Death does not involve the corruption of Socrates, but Socrates's body. Again, this passage offers explicit support of the alternative account of Socrates's death and, along with it, the view that the substantial form of Socrates is necessary and sufficient to preserve Socrates's identity through time and change. Aquinas's understanding of the death and resurrection of human beings thus offers us additional grounds for thinking that he accepts AI instead of AI*.

Aquinas on the individuation of material substances

We have seen that Aquinas takes the substantial form of a material substance to be the necessary and sufficient condition for that material substance remaining numerically the same through time and change. Now I turn to the question of explaining what individuates – that is makes individual – the substantial form of a material substance and, with that form, the material substance itself. The question as to what ultimately explains why a material substance is an individual distinct from others is often referred to as the 'problem of individuation'.

Designated matter as principle of individuation

For Aquinas 'the problem of individuation' is a question that primarily has to do with material substances. To see why, note that, unlike material substances, immaterial substances do not share a species according to Aquinas.[28] Whereas Socrates and Plato are one in substance in the sense that they both belong to the species *human being*, Michael and Gabriel are not one in an analogous sense. For Aquinas, every immaterial substance is a species of being unto itself – every angel is as it were, 'the only one of its kind'.[29] Of course, such an account of the angels raises questions of its own, questions which I can't address here. But Aquinas does think that Socrates and Plato do share a species and so they have in common everything that all human beings have in common, e.g. having a human nature, both biological and psychological. But with all that Socrates and Plato have in common the question remains, what is it about Socrates and what is it about Plato that explains the fact that they are numerically distinct from one another? Aquinas clearly holds throughout his career that neither matter, nor substantial forms, nor accidents can by themselves explain why Socrates and Plato differ in number. Rather, it is 'designated matter' – matter that can be pointed at with the finger – that provides the ultimate explanation for the individuation of material substances.

Why can't form of some sort be the principle of individuation? Aquinas offers the following argument: material forms – forms that exist extramentally only in so far as they are received in matter – are, considered in themselves, communicable to many things. But whatever is capable of being received into many is 'contrary to the nature of that which is a *this something*'.[30] So neither substantial forms nor accidental forms can serve to individuate Plato and Socrates in the species *human being*.

As we've seen, Aquinas thinks that the substantial form of a material substance explains why that substance belongs to the species that it does. Such a thought seems readily compatible with what Aquinas says about the inability of substantial forms to individuate a material substance. However, as I noted in Chapter 4, insofar as they configure matter substantial forms (and accidents as well) are individuals and not universals. Given that substantial forms (and accidents) only exist extra-mentally as individual realities, why can't they then serve the purpose of individuating a material substance? In that case Socrates would be distinct from Plato in that Socrates has *this* substantial form, *this* whiteness, etc., whereas Plato has *that* substantial form, *that* whiteness, etc. Socrates and Plato would be different human beings insofar as they have different substantial and/or accidental forms.

But Aquinas thinks that material forms – forms that are real only insofar as they are received in matter – are not individuals *unless* they are received in matter. Substantial forms too are *individuated*, and they are individuated as a result of being received in matter. Saying that two material things have different forms seems to presuppose some further difference in the subjects that receive such configurations. As Brian Davies writes,

> It may ... be true that *A* is different from *B* because *A* is short, pale and abrasive while *B* is tall, tanned and placid. But ... in pointing to such differences as differences between *A* and *B* we are already presupposing that *A* and *B* are distinct.[31]

In other words, in pointing to formal differences between Socrates and Plato in order to individuate them, one is already assuming that Socrates and

Plato are two different objects such that they can have different substantial and accidental forms. Why think there are two different objects present? Aquinas's answer: because Socrates is *here* whereas Plato is *there*.

It might seem therefore that matter is the principle of individuation for Aquinas and, given the right sort of clarifications, this is indeed the position that Aquinas takes with respect to the problem of individuation. But it should be clear why *prime* matter itself couldn't function as principle of individuation for Aquinas. This is because prime matter itself is not something undifferentiated – since it is matter without any form. As Aquinas notes, considered by itself, that is, apart from a configuring substantial form, prime matter is one in number.[32] Prime matter is not one in number because it has a form that makes it *this* thing rather than *that* thing, but because it is completely lacking in all form. Prime matter can't explain why Socrates and Plato are one in species but different in number.

The matter that explains why Socrates and Plato are individuals within the same species must itself be something individuated.[33] How is Socrates's matter different from Plato's? Aquinas follows Aristotle in saying that prime matter exists as something divisible and distinct in virtue of quantity.[34] Aquinas thus concludes that matter is made *this* matter – what Aquinas calls '*materia signata*' (designated matter), or matter that can be pointed at with the finger – insofar as it has dimensions.[35] Again, it is because Socrates has matter *here* and Plato has matter *there* that it makes sense to say that Plato and Socrates are not one and the same human being, but rather distinct human beings.

Aquinas sometimes adds a further detail to his view that matter is the principle of individuation. The dimensions of a material substance can be considered as either terminated or un-terminated.[36] If the dimensions of some material substance x are considered as terminated, then those dimensions are being considered according to the precise figure that x has at some time t.[37] Of course, the terminated dimensions of a material substance x are constantly changing from moment to moment so that if matter under *terminated* dimensions were to function as the principle of individuation, then x wouldn't enjoy an existence longer than a few seconds.[38] Nevertheless, we can also consider the dimensions of a material substance x as un-terminated. A material substance x is by its very nature something spread out in space – having dimension – and therefore x excludes any substance not-x from being where x is at any given time. To observe that two things cannot exist in the same space at the same time, without concern for the precise (terminated) dimensions of those two things, is to acknowledge that matter can be considered simply insofar as it is spread out in different spaces, that is, that different things have different matter under un-terminated dimensions. It is matter considered in this way – as un-terminated – that Aquinas takes to be the *principle* of individuation.[39]

Aquinas on the individuation of non-living substances

I now would like to consider how Aquinas applies his theory of individuation to some difficult cases. It is easy to see how Aquinas's theory applies to cases involving living substances such as Rex and Fido. Rex and Fido are individual dogs insofar as they exist as extended in space in different places. Furthermore, if Rex loses his tail in an accident, it is not hard to see that Rex does not go where his tail goes. Rather, the tail-less entity has the same matter under un-terminated dimensions as Rex does with a tail. But consider the case of a non-living substance such as a puddle of water that is divided into two. Is either of these puddles after the division identical to the original puddle?

There is an interesting passage in Aquinas's ST where he speaks about a homeomerous substance undergoing a variety of changes. His example treats a case involving fire:

> As the Philosopher says in I *De generatione*, when some matter is converted into fire, it is now called a newly generated fire, but when some matter is converted into a pre-existing fire, it is said that the fire is fed. Hence, if the whole matter loses the species of fire all at once, and another matter is converted into fire, that fire will be a fire other in number. But if little by little as one piece of wood is burned another is substituted, and so forth, until the first piece is consumed entirely, the fire will remain always the same in number since what is added always passes into what is pre-existing.[40]

This passage suggests that, like a living substance, a non-living substance can retain its identity through time and change despite undergoing changes in its terminated dimensions. So non-living substances can 'grow' (and presumably shrink) without going out of existence. This is compatible with Aquinas's view that matter under un-terminated dimensions is the principle of individuation.

But what about a case where a non-living substance is suddenly divided into two or more distinct things? Say there is a puddle of pure water (call it 'Puddle') in front of my car at time *t*, and at *t* + 1 I step in Puddle with such force that it is split into two puddles of water (call these 'Puddle1' and 'Puddle2', respectively). It would seem to be rather ad hoc to identify Puddle with either Puddle1 or Puddle2. In contrast to cases involving the division of a heterogeneous substance such as a dog, where it is clear that some parts of the substance simply cannot embody the substantial form of the substance in question, e.g. a dog's tail can't be a dog, things are not so clear with home-omerous substances such as masses of water. Say that Puddle1 is larger than Puddle2. Perhaps Puddle should be identified with Puddle1 since it is larger. But why think Puddle should be identified with the larger (or smaller) of the two puddles? It makes more sense to say that Puddle simply goes out of existence whenever it is divided, so that Puddle is not identical to either Puddle1 or Puddle2.[41]

This appears to be the position that Aquinas in fact takes. In one place he argues that some material forms are 'extended along with matter', while others, namely those of the higher animals, are not. Aquinas appears to be tying a substance having its form extended in matter to its being non-heterogeneous.[42] Aquinas gives a worm as an example of a material substance whose substantial form is 'extended along with matter'. He suggests therefore that the soul of a worm is actually one, but potentially many. In contrast to a higher animal such as a dog, the soul of a worm – like the substantial form of a portion of water – is a form that *is* extended along with matter, and all forms that are extended in matter can be divided.[43] But the division of a substantial form surely brings about that substantial form's dissolution, and with it the corruption of the substance whose substantial form it was. If a worm is divided – at least where the result of the division is two worms – neither of the worms that result from that division are identical with the original worm. Guided by this story about the division of a worm's soul, I want to propose *a fortiori* that Aquinas has the view that *any* division of a non-living substance has as its result that the divided substance goes out of existence.[44]

Aquinas on the individuation of human beings

As we saw in Chapter 3, Aquinas thinks that the substantial forms of human beings are incorruptible, and so specially created by God. This raises questions about whether designated matter is in fact the principle of individuation for human beings. If human souls were not individuated by matter, this would raise a concern about whether human beings are material things after all. Since, as we have seen, Aquinas does think that human beings are material things, if human souls were individuated by matter, then Aquinas's views on human beings would be logically inconsistent.

Indeed, a powerful argument can be put forward to the effect that for Aquinas the human soul is not individuated by matter, but is rather something individual through itself. Call the argument 'IH':

(1) Everything is individuated in the same way in which it has being. (2) Human souls do not have their being in virtue of being united to matter, but *per se*. First proof of (2): human souls can exist apart from matter. Second proof of (2): human souls are not educed from matter as are material substantial forms; they are specially created by God. (3) Therefore, human souls are not individuated by matter but through themselves.

First of all, it is clear that Aquinas rejects the conclusion of IH. For example, in discussing a question about whether human beings can have different degrees of knowledge about the same thing, Aquinas entertains the following objection: if different human beings could have different degrees of knowledge about the same thing, then human intellects would differ from one another. But human intellects are substantial forms, and substantial forms that differ from one another differ by species. Since human souls do not differ by species, they do not have different degrees of knowledge about the same thing.[45] Aquinas answers the objection above by saying that if x and y differ on account of having different matter, then x and y differ numerically and not specifically. But human intellects differ on account of matter, and therefore differences in the intellects of x and y do not cause a specific difference between x and y but only a numerical difference.[46] Thus, here we have Aquinas specifically saying that the substantial form of a human being (which Aquinas refers to simply as 'intellect' in the passage I have been examining) is individuated by matter.[47] Since Aquinas rejects the conclusion of IH, and IH is logically valid, he must reject one of IH's premises if his views are to be logically consistent.

But Aquinas clearly accepts premise 1 of IH. Something is individuated in the same way that it has unity. But unity and being are convertible. Therefore, something is individuated in the same manner in which it has being.[48]

The problematic premise of IH is (2) as far as Aquinas is concerned. The problem with (2) is that it does not say enough about the *nature* of the existence of human souls. It is true for Aquinas that human souls can exist apart from matter. But, as we saw in Chapters 3 and 4, Aquinas also thinks that human souls are substantial forms of some body *by their nature*. Thus, I think Aquinas would replace (2) of IH with the following premise:

(2*) Human souls, in accord with their being, are united to a body as their form; nevertheless, upon the destruction of the body, intellectual souls retain their own being.[49]

What conclusion follows from the conjunction of 1 and 2*? Rather than concluding that human souls are not individuated by matter *simpliciter* as in (3), one could conclude that human souls are individuated through their coming to configure some matter – as is the case with *material* substantial forms – but in a way that does not prevent those souls, once they exist (and are therefore individuated), from existing on their own apart from matter. Indeed, this is the picture of the being and individuation of the human soul that Aquinas presents throughout his corpus, although he expresses this view in a variety of different ways.[50] The soul is indeed individuated by matter, since the soul is created by God as a substantial form that configures matter by its nature; but once the soul is individuated by matter in its origins, it no longer depends on matter for its individuation. As Aquinas puts it one place, 'the individuation and multiplication of [rational] souls depends upon the body with respect to its beginning, but not with respect to its end'.[51] Some corollaries of the view that the soul's individuation has its origin in matter include Aquinas's views that the soul *begins* to exist as a form configuring matter – the soul cannot pre-exist the matter it configures[52] – and once the soul configures a particular body, it cannot configure a numerically different one.[53]

The view that the soul is individuated through its relation to matter, but not in a way that prevents its existing apart from matter once it is individuated, is in fact what we would expect Aquinas to say given his understanding of the ontological status of the human soul. The human soul is not a substance; nor is it a garden-variety substantial form. Rather, it is a *subsistent* substantial form. Because it is a substantial form, it is individuated by matter. But because the human soul is subsistent and can exist without the matter that it naturally configures, it is not individuated by matter in the same way that non-human material substances are.

Aquinas on the identity and individuation of artefacts

Having explained Aquinas's views on the identity and individuation of material substances, I now want to turn to Aquinas's views on the identity and individuation of artefacts.[54] As we saw in Chapter 4, Aquinas thinks of artefacts as composites of matter and form. However, instead of being a composite of a substantial form and prime matter, an artefact is a composite of an actually existing substance (or aggregate of substances) and an accidental form. One of the implications of this view is that artefacts are not substances themselves. In this sense, Aquinas has a view of artefacts that parallels a reductive account of compound material substances.

Aquinas on the numerical identity of artefacts through time and change

Since artefacts are like substances in being composites of form and matter, one might expect that a Thomistic account of the identity of artefacts is roughly parallel to the Thomistic account of the identity of substances. Since Aquinas thinks that material substances x and y are identical if, and only if, x and y have the same substantial form, we would expect the following account of artefact identity:

(IA) For any artefacts x and y, x is numerically identical to y if and only if x and y have the same artefact-configuring accidental form.[55]

IA does, in fact, capture Aquinas's views on artefact identity. First of all, Aquinas takes sameness of artefact-configuring form to be a *necessary* condition for an artefact's preserving its numerical identity. Consider the following passage:

> A statue can be considered in two ways, either insofar as it is a certain substance, or insofar as it is a certain artificial thing. And since a statue is put into the genus of substance by reason of its matter, therefore, if it is considered insofar as it is a certain substance, the statue that is restored through the same matter is the same in number. But if a statue is put into the genus of artificial things insofar as it has a form that is a certain accident, and [the statue] loses [that form], the statue [will be] destroyed. Thus, [the form] does not return the same in number, and neither is the statue able to be the same in number.[56]

Note first that a statue can be considered in two ways: as a substance and as an artefact. Now a statue is not considered as a substance because an artefact really is a substance. Rather, as Aquinas makes clear in the passage, a statue (an artefact) is considered a substance only in virtue of its matter, e.g. some bronze. A bronze statue is considered a substance insofar as it is constituted by some bronze. Thus, it is really the matter of a statue and not the statue itself which is a substance. More important for our immediate purposes, Aquinas says that it is because a statue's artefact-configuring form is lost that that statue goes out of existence. Indeed, a statue loses its identity upon losing its artefact-configuring form even if the matter of the statue (e.g. some bronze) remains numerically the same through the change. Here Aquinas definitely shows that he thinks having the same artefact-configuring form is a necessary condition for artefact identity.

But the passage we have been examining also seems to suggest that having the same artefact-configuring form is a *sufficient* condition for artefact identity given Aquinas's line, 'a statue is put into the genus of artificial things insofar as it has a form that is a certain accident'. This makes sense because accidental forms are like material substantial forms in that they cannot exist apart from matter, and (as we shall see in the sequel) accidental forms are individuated by the substances that they modify. Thus, Aquinas would appear to accept the following:

> (AF) If artefacts x and y have the same artefact-configuring form, then it is the case that artefacts x and y have the same matter.

Of course, artefacts that have the same form and matter are obviously identical. Since (assuming the correctness of AF) it is the case that artefacts that have the same artefact-configuring form therefore have the same matter, if artefacts x and y have the same artefact-configuring form, then artefacts x and y are numerically identical.

Thus, Aquinas does accept IA. Sameness of artefact-configuring form is a necessary and sufficient condition for the identity of an artefact through time and change.

Aquinas on the individuation of artefact-configuring forms

According to AF, sameness of matter is a necessary condition for sameness of an artefact's artefact-configuring form. This view is itself a specific application of a general principle that Aquinas accepts, namely, that accidental forms are individuated by the substances they modify. In fact, that accidents are individuated by substances is itself an application of an even more general principle that Aquinas accepts: forms configuring matter are individuated by the matter that they configure. Just as Aquinas thinks that matter (understood in a certain way) is the principle of individuation for material substances, so too matter (understood in a certain way) is the principle of individuation for accidents. Of course, accidents do not have a material cause in the same way that material substances do. But accidents are received by something, namely, the substances that they modify. And it is what receives form that is primarily responsible for its individuation. Thus, Aquinas thinks that substances individuate the accidents that modify them.[57]

However, notice that artefacts x and y having the same matter is only a necessary condition for the identity of artefact-configuring forms x and y. It is true that for any artefact-configuring forms x and y at any individual time t, x and y are numerically identical if the matter of x and y is the same at t. But if we are talking about the identity of artefact forms *across* time, identity of matter is not a sufficient condition for the identity of artefact forms. This is because the same substance might be modified by the same species of accident at two different times, with a temporal gap between these two modifications. For example, Jane was tanned last summer, and she is also tanned this summer. But since she was very pale during the intervening winter season, in Aquinas's view, Jane's tan this summer is not the same as her tan last summer.[58] As we have seen, Aquinas does not think that forms or substances can intermittently exist. Hence, accidental forms x and y are the same only if x and y have existed at the very same times. But artefact-configuring forms are accidental forms. Therefore, artefact-configuring forms x and y are the same only if x and y have existed at the very same times.

Aquinas mentions at least one other necessary condition for the sameness of artefact-configuring forms x and y: x and y are the same only if x and y are the numerically same form of *order* and/or *composition*.[59] Permit me to explain by way of an example. Suppose Jane's house is a split-level and is composed of a certain aggregate of pieces of wood at time t (call the pieces of wood that compose Jane's house at t 'the xs'). Now Jane decides she wants to change the style of her house to that of a ranch. However, since Jane is short on funds, she makes use of the same pieces of wood that compose her split-level in constructing the ranch-style house. Say the remodelling project is completed at time $t+1$. Now the remodelled ranch-style house has a different configuration from Jane's old split-level, although this new configuration was brought about without ever turning the *x*s into a non-house (Jane made use of her house in the same ways she always did throughout the remodelling process). Furthermore, the split-level and the ranch are both composed of the *x*s. As Aquinas makes clear in one text, he thinks that artefacts with different kinds of configurations – having forms that differ with respect to order and/ or composition – have different artefact-configuring forms.[60] Now Jane's split-level and Jane's ranch are both composed of the *x*s (the houses are composed of the same matter) and Jane never lacked a house between t and $t+1$ (there is at least an *appearance* of continuity of artefact-configuring form between t and $t+1$). But Jane's house at t has a kind of configuration different from Jane's house at $t+1$ and thus Jane's house at t and Jane's house at $t+1$ have different artefact-configuring forms. Given Aquinas's account of artefact identity, that means that Jane's house at t is not numerically the same house as Jane's house at $t+1$.

Implications of Aquinas's views on artefact identity

Aquinas's views on artefact identity and individuation suggest that he has a very restrictive notion of artefact identity over time. According to Aquinas, artefacts *x* and *y* are numerically identical if, and only if, *x* and *y* have numerically the same artefact-configuring form. But in order for artefacts *x* and *y* to have numerically the same artefact-configuring form, the artefact-configuring forms of *x* and *y* must:

(a)   configure the numerically same matter, and

(b)   exist at all the same times.

Furthermore, *x* and *y* are the same only if

(c)   *x* and *y* are specifically the same form of order and/or composition.

Consider the implications of such criteria for artefact identity by first taking a look at a case involving a simple artefact, e.g. a statue constituted by a piece of clay. According to Aquinas, such a statue is a composite of a piece of clay (a substance for Aquinas) and an artefact-configuring form (the particular shape of the clay that makes that clay a statue). Suppose the statue is broken so that a small part of the statue is *divided* from the statue as a whole. Given

Aquinas's views on the identity and individuation of non-living substances, the clay that functions as the matter of the statue goes out of existence when it loses a part through a process of division, becoming a numerically different substance of the species *clay*. But this means that the statue goes out of existence as well, because condition (a) for artefact identity is no longer met. That is, there may be a statue present after the original statue loses a part in virtue of the clay's being divided but, if there is a statue present at that time, that statue is not numerically the same as the original statue. Consider instead that the clay of the original statue is still a little wet, and instead of the original statue being broken, it has its shape altered slightly. It would appear that the statue that results (one that is bent) is not numerically the same as the original statue, since in such a case condition (c) is violated. The bent statue has a shape or composition different in kind from the original statue and so the bent statue and the original statue would seem to be numerically different statues by Aquinas's lights.

Now take a case of change involving a *complex* artefact such as a log cabin. (Recall that a 'complex' artefact – in contrast to a 'simple' artefact – has an aggregate of material substances functioning as its matter.) To make the case easy, let us say that this log cabin is very primitive; its material quotient is simply an aggregate of logs. According to Aquinas, the log cabin will also have an artefact-configuring form, e.g. a certain ordering of the logs. Now it seems as though one can alter the numerical identity of the log cabin simply by replacing *one* of the logs of the cabin. For aggregates are individuated by their members, and an aggregate of logs with logs *a*, *b*, *c*, *d* and *e* is an aggregate different from one having logs *a*, *b*, *c*, *d* and *f* as members. So replacing a log in the cabin would appear to violate condition (a) for artefact identity. That would mean that one could destroy a log cabin simply by replacing *one* of its logs. Say instead that one takes apart the cabin, and then proceeds to put the same set of logs back in precisely the same order. Given Aquinas's criteria for artefact identity, the reconstructed cabin will not be the same cabin as the original, for though the artefact-configuring forms of the reconstructed and original cabins may be specifically the same, they cannot be numerically the same given that condition (b) is violated: the artefact-configuring form of the original cabin went out of existence when the original cabin was dismantled. Since Aquinas rejects the possibility of intermittent existence, the artefact-configuring form of the reconstructed cabin can't be the same as that of the original. It also appears that the moving of any of the logs would also cause the cabin to go out of existence. For in such a case, condition (c) would be violated – the composition of the integral parts of the cabin having been altered.

Now these conditions for artefact identity may seem (excessively) restrictive. After all, if I break off one of the keys of my computer, the computer that I continue to use would seem to remain the very same computer that I purchased from the local discount electronics store last year (at least, the warranty agreement says so). Indeed, many contemporary philosophers have expressed an allegiance to the following intuition about artefact identity (or 'IAI' for short):

(IAI) Although artefacts cannot survive a complete replacement of their parts, they can survive a number of different partial replacements of their parts, or alterations in their component parts, even if it is admittedly unclear precisely to what extent an artefact can undergo part replacements, or alteration of its parts, and still remain numerically the same artefact.[61]

Aquinas's views on artefact identity are incompatible with IAI. Of course, the reader should note that one of the reasons for this is that Aquinas thinks that artefacts are things composed of *non-living substances*.[62] And as we saw above, Aquinas thinks that non-living substances are extremely fragile. Because the identity of an artefact comes and goes with its artefact-configuring form, and the identity of the artefact-configuring form has sameness of configured matter as a necessary condition, artefacts having such substances as their material quotient are going to be very fragile as well.

Before judging Aquinas's restrictive view of artefact identity too harshly, the reader should keep in mind that artefacts are not substances for Aquinas. His views about artefact identity should be seen in such a light. Authors such as Lynne Rudder Baker who would reject the restrictive Thomistic account of artefact identity on the grounds that it violates intuitions about artefacts would also, of course, reject the theoretically more basic view of Aquinas's that artefacts do *not* enjoy the same ontological status as material objects such as pieces of clay, trees and frogs. It makes sense for Aquinas to have restrictive criteria for artefact identity in light of his views on the ontological status of artefacts, namely that they are not substances. One might put this point in another way: those who find Aquinas's restrictive view of artefact identity absurd are probably already committed to accepting the intuition that artefacts are on an ontological par with living things. And *that* intuition is certainly a questionable one.

But perhaps what I am calling Aquinas's *restrictive view* of artefact identity – based as it is on only a few texts – can be softened a bit in light of Aquinas's admittedly more detailed discussion of the individuation of material substances. Specifically, the restrictive view might be somewhat amended, bringing it closer to IAI. Now IAI requires that an artefact be able to survive any small change with respect to its component parts. What I want to suggest here is that conditions (a) and (c) of the restrictive view of artefact identity might be weakened in ways that find them more in accord with IAI.

But first I want to say something else about condition (b). Although IAI has nothing to say about condition (b), many authors who find IAI compelling will also be committed to the possibility of intermittent existence, particularly with respect to artefacts.[63] Take the following example. Joe is a very curious boy with a bicycle. He wants to know how the bike works, and so (in good reductionist fashion) reasons that to find out something about the nature of the bike, he should take it apart and examine each of its parts individually. He takes apart the bike completely. Eventually his desire to ride the bike again overcomes his desire for theoretical knowledge, and so he puts the bike back together and, as it happens, puts the bike back together so that the gears, spokes – everything that was part of the original bike – is precisely in the same place on the bike as it was before. A question: is Joe now riding the numerically same bike as before? Someone might be inclined to answer, 'of course'. But, according to Aquinas's restrictive view, the right answer is 'no'. Anyone who believes that the gears, the handlebars, etc. can indeed compose a bike also thinks that the gears, handlebars, etc., when they are lying on the ground so that they are in no way contiguous, do not at that time compose a bike. This means that, if the reconstructed bike is numerically the same as the original, then Joe's bike existed for a time, went out of existence, and then came back into existence – a case of intermittent existence. As I have argued, Aquinas is committed to the impossibility of intermittent existence. Note also that condition (b) can't be given a weaker interpretation. A form is either able to exist intermittently or it is not. Thus, a weaker interpretation of Aquinas's conditions for artefact identity can't include a weakening of condition (b).

In contrast, conditions (a) and (c) for artefact identity can be weakened, I think. What I want to suggest here is a loosening of these conditions in much the same way that Aquinas does when he talks about the dimensions of material *substances*. As we have seen, Aquinas thinks that the principle of individuation in material substances is designated matter, or matter under un-terminated dimensions. Aquinas posits that matter under *un-terminated* dimensions is the principle of individuation rather than matter under terminated dimensions since the terminated dimensions of virtually all material substances are constantly changing, and Aquinas takes it as an obvious truth that at least some material substances can survive changes in their terminated dimensions. Given that Aquinas leans upon intuitions with respect to the compatibility of identity and mutability in material substances, why could he not soften the restrictive view of artefact identity in much the same way?

Let me explain what I am proposing here by returning to some of the examples I used above. I said that according to Aquinas's restrictive view, the log cabin goes out of existence if any of the logs in that cabin are *replaced* – such situations would involve violations of condition (a) – or *moved* – this would involve a violation of condition (c). But in the same spirit of allowing for fluc-tuations in the dimensions of material substances throughout their careers, I want to suggest that some parts of an artefact are essential to it, and others are not. Any log cabin needs a roof and four walls. Insofar as a log cabin's roof and walls remain in existence, that log cabin remains in existence. Now a roof with a hole in it is nonetheless still a roof. Since a wall can survive the loss of a part, the log cabin composed of that wall can survive the loss or replacement of a log in that wall. Analogously, certain compositional modifications will be essen-tial to a particular log cabin, while others will be accidental. However, notice that even on IAI, it may not be completely clear which features of a log cabin are essential to it, and which are accidental.

It is more difficult to see how to adjust Aquinas's restrictive view in the case of a simple artefact such as a clay statue when that statue either loses a part or undergoes an alteration of shape. According to the restrictive view, a clay statue would go out of existence if the substance that functions as the matter of the statue were to suffer any sort of division (a violation of (a) – since when the matter of the statue goes out of existence upon being divided, so does its accompanying artefact-configuring form) or the shape of the artefact were in any way altered (a violation of condition (c) – the order of the clay having been changed). I think it is easy to see how one might offer a weaker interpre-tation of condition (c). The clay statue is a statue and not simply a lump of clay because it has a certain kind of shape. Not any old shape that a piece of clay has entails that it is the matter of a statue according to Aquinas. Aquinas speaks in one place about a piece of bronze before it becomes the matter of a statue as something 'unshaped or unarranged' (*unfiguratum sive indispositum*).[64] But Aquinas can't mean by the locution 'unshaped bronze' that the described piece of bronze has no shape at all. Every material substance has some sort of accidental form in the genus 'shape'; *non-living* material substances simply do not have to have any one particular shape. Obviously, the sort of shape Aqui-nas has in mind is that of a figure, say, a representation of a great Hebrew king as in Michelangelo's *David*. Of course, a statue obviously can't survive all changes in shape. A clay statue doesn't survive being melted down. But what I am suggesting here is that artefacts such as statues – which are statues in virtue of having a certain kind of shape – are such that their shapes have essential and accidental features.

As the case of the *Venus de Milo* makes clear, we commonly think that a statue can lose some of its parts, and still remain numerically the same figure. But it looks difficult to see how Aquinas's views can accommodate this intui-tion. A simple artefact (such as a statue made of one piece of material) by defi-nition has one material substance as its material component according to Aquinas. But in Aquinas's view, a non-living material substance goes out of existence whenever it is divided. Since having the same matter is a necessary condition for the preservation of the identity of an artefact-configuring form, a simple artefact is corrupted whenever the matter that composes it is divided. In the case of a simple artefact such as a statue one cannot soften condition (a) by suggesting that some members of the aggregate that form the material component of the artefact in question are essential to it while others are not. In the case of a simple artefact (such as a clay statue) there is only *one* substance that functions as its material component. There seems to be no room for manoeuvring here.

A reader might offer the following solution to this objection. Why not weaken the conditions for non-living material substances along the same lines as the conditions for complex artefacts? In other words, why not say that non-living substances (which function as the material quotient of a simple artefact), like complex artefacts, have essential and non-essential parts? But there is this important difference between complex artefacts and non-living substances that makes this suggestion untenable: complex artefacts are heterogeneous in nature whereas non-living substances are not. Since a complex artefact is heterogeneous in nature, it makes sense to say that some of its parts are essential to its identity whereas others are not, e.g. a house must have a roof as a part but it need not have a deck. But the non-heterogeneous nature of non-living substances prevents us from applying the distinction between essential and non-essential parts to such substances. If a non-heterogeneous substance is divided, how could either of the resulting halves be the same as the original? Indeed, any explanation as to why one (former) part is identical to the whole would be ad hoc. Therefore, since a non-living substance can't survive a division, a simple artefact has a non-living substance as its material quotient, and the identity of a simple artefact has sameness of matter as a necessary condition, condition (a) cannot be weakened for simple artefacts in the way that it can for complex artefacts.

However, there is a sense in which *all* artefacts are (or, at least, appear to be) more like living substances than non-living substances. For even simple artefacts are heterogeneous objects, e.g. a human statue made of clay has different parts that represent the different parts of the human body. Thus, whether or not Aquinas's restrictive view of artefact identity can or can't be modified in the ways I have suggested here, it is at least clear why we intuitively think (perhaps mistakenly) that artefacts can survive the same sort of changes that living substances can: like living substances, artefacts are heterogeneous objects.

Notes

1. See also my discussion of relative identity in Chapter 1.
2. Stump 2003, p. 45.

3. See, e.g., Shoemaker, S. (1998) 'Personal identity: a materialist account' in P. van Inwagen and D. Zimmerman (eds) *Metaphysics: The Big Questions*, Malden: Blackwell, pp. 298 and Hirsch, E. (1995) 'Identity' in J. Kim and E. Sosa (eds) *A Companion to Metaphysics*, Oxford: Blackwell, p. 230.

4. See, e.g., Lewis, D. (1998) 'The problem of temporary intrinsics: an excerpt from *On the Plurality of Worlds*' in P. van Inwagen and D. Zimmerman (eds) *Metaphysics: The Big Questions*, Malden: Blackwell, pp. 204–5 and Zimmerman, D. (1998) 'Temporary intrinsics and presentism', in P. van Inwagen and D. Zimmerman (eds) *Metaphysics: The Big Questions*, Malden: Blackwell, pp. 206–19. My own discussion of these issues is indebted to Stump 2003, pp. 44–6.

5. It may be that I am wrong that LII is ambiguous in the way that I have been suggesting, that it is obvious that 'identical' in LII has to be understood as *numerical* identity. In that case, LII relativism simply amounts to a rejection of LII. Again, the advocate of LII relativism would think this reasonable given our common-sense understanding of change over time, one that suggests that an object's remaining numerically the same object is compatible with that object having different (accidental) properties at different times.

6. See, e.g.: In Met. V, lec. 11, nn. 907 and 912 and In Met. V, lec. 17, n. 1022.

7. For Aquinas '*x* is identical to *y*' and '*x* is the same as *y*' are logically equivalent expressions (see, e.g.: In Met. V, lec. 11, nn. 907, 911 and 912; In Met. X, lec. 4, n. 2007).

8. Of course, accidents too can be regarded as one in a sense, but they are one in a sense that is only analogous to how substances are one according to Aquinas. For discussion of how accidents are one, see: In Met. V, lec. 11, nn. 907 and 912; In Met. V, lec. 17, n. 1022, and In Met. X, lec. 4, nn. 2007–8.

9. See, e.g.: In Met. X, lec. 4, nn. 2001–5 and In Met. V, lec. 11, n. 912. See also ST Ia. q. 11, a. 1, ad2, for a similar list.

10. See, e.g., Stump, E. and Kretzmann, N. (1985) 'Absolute simplicity', *Faith and Philosophy* 2: 369–70.

11. See, e.g., In Met. V, lec. 8, n. 880.

12. See, e.g.: ST Ia. q. 119, a. 1, ad5 and DGC I, lec. 16, n. 112.

13. Stump 2003, pp. 44–6. For texts in Aquinas that explicitly support AI, see, e.g.: In Sent. IV, d. 44, q. 1, a. 1, sol. 2, ad4 (ST Suppl., q. 79, a. 2 ad4); QQ 11. q. 6, a. 1, ad3; SCG II ch. 65 (4); SCG II ch. 83 (37); SCG IV ch. 81 (6, 10, 11, 12); In Met. VII, lec. 13, n. 1588 and ST Ia. q. 119, a. 1, ad5.

14. See, e.g.: SCG IV ch. 80 (2), SCG IV ch. 81 (5–11), and CT ch. 153.

15. See, e.g., ST Ia. q. 75, aa. 2 and 6.

16. See, e.g., ST Ia. q. 75, a. 4.

17. I learned what I am calling the *alternative account* of Socrates's ontological status after death and before the general resurrection from Eleonore Stump. See 2003, pp. 51–4.

18. Robert Pasnau has recently defended an interpretation of Aquinas's views on the issues I am discussing here that neither conforms strictly to the corruptionist account nor the alternative account I present and defend here. See Pasnau, R. (2002) *Thomas Aquinas on Human Nature*, Cambridge: Cambridge University

Press, pp. 379–93. Pasnau's view is that Socrates does survive his biological death, but in an ontologically deficient state. That is, Socrates enjoys a 'partial' existence after his death and before the general resurrection. Nevertheless, Pasnau's conception of a human being partially existing after death and before the resurrection is itself close to – although in my view not as perspicuous as – the alternative account I defend here. In critically examining Pasnau's views on Aquinas and the resurrection, I benefited greatly from conversations with Jason T. Eberl.

19. See, e.g.: SCG II ch. 65 (4); SCG II ch. 83 (37); SCG IV ch. 81 (6, 10, 11, 12); In VII Meta., lec. 13, n. 1588 and ST Ia. q. 119, a. 1.

20. See also my discussion of subsistent substantial forms in Chapter 4.

21. Aquinas has at least two reasons for taking this latter view. First, Aquinas takes 'resurrection' to mean the assumption of life in the same body that one had before one's biological death. If the soul were joined to a body different from the body possessed during one's life, this would not be a case of resurrection, but (perhaps) a reincarnation. See, e.g., In Sent. IV, d. 44, q. 1, a. 1, sol. 1 (ST Suppl. q. 79, a. 1, c.). Second, Aquinas has the general view that substantial forms have an appropriate corresponding matter. See, e.g., SCG II chs 80–1 (7). For some good discussion of Aquinas's adherence to this principle, see Pasnau 2002, pp. 380–91.

22. See, e.g.: ST Ia. q. 2, a. 1, c.; ST Ia. q. 13, a. 12, c.; ST Ia. q. 76, a. 3, c. and ST Ia. q. 76, a. 3, ob. 4. See also SCG II chs 57 (5) and 58 (3 and 7), where Aquinas says that humans are animals essentially.

23. Stump discusses other problems for what I am calling the 'corruptionist account': it cannot explain the fact that Aquinas attributes distinctively human properties to the separated soul, e.g. having beliefs and desires, the ability to understand, will, and intercede in prayer for other human beings, etc. See 2003, p. 52.

24. SCG IV ch. 80 (3): 'Quod non est continuum, idem numero esse non videtur. Quod quidem non solum in magnitudinibus et motibus manifestum est, sed etiam in qualitatibus et formis: si enim post sanitatem aliquis infirmatus, iterato sanetur, non redibit eadem sanitas numero. Manifestum est autem quod per mortem esse hominis aufertur: cum corruptio sit mutatio de esse in non esse. Impossibile est igitur quod esse hominis idem numero reiteretur. Neque igitur erit idem homo numero: quae enim sunt eadem numero, secundum esse sunt idem.'

25. SCG IV ch. 81 (11): 'Manifestum est enim quod materiae et formae unum est esse: non enim materia habet esse in actu nisi per formam. Differt tamen quantum ad hoc anima rationalis ab aliis formis. Nam esse aliarum formarum non est nisi in concretione ad materiam: non enim exedunt materiam neque in esse, neque in operari. Anima vero rationalis, manifestum est quod excedit materiam in operari: habet enim aliquam operationem absque participatione organi corporalis, scilicet intelligere. Unde et esse suum non est solum in concretione ad materiam. Esse igitur eius, quod erat compositi, manet in ipsa corpore dissoluto: et reparato corpore in resurrectione, in idem esse reducitur quod remansit in anima.' See also In Sent. IV, d. 44, a. 1, sol. 1, ad4 and In Sent. IV, d. 44, a. 1, sol. 2, ad 1.

26. Other passages in Aquinas's corpus also confirm his rejection of IE. See, e.g.: ST Suppl., q. 79, a. 2, ad4; SCG II ch. 83 (37), and In Phys. V, lec. 6, n. 700.

27. In this connection, see also the following texts: CT, ch. 154, where Aquinas makes it clear that he thinks that it is the restoration of the (numerically same) body and its various organs at the resurrection that requires a divine miracle and not the restoration (re-creation) of an entire human being, and SCG IV. ch. 81, sections (6)–(10), where Aquinas explains the possibility of Socrates retaining sameness of body at the resurrection by appealing to the permanence of *his* soul and not by way of divine re-creation.

28. There is a technicality in Aquinas's views that I am overlooking here, namely that some material substances are not in fact divided within a species either, e.g. the heavenly bodies. See, e.g. ST Ia. q. 66, a. 3, c.

29. See, e.g., DEE ch. 5.

30. In BDT q. 4, a. 2, c.: 'Intellectus enim quamlibet formam, quam possibile est recipi in aliquo sicut in materia vel in subiecto, natus est attribuere pluribus, quod est contra rationem eius quod est hoc aliquid.'

31. Davies, B. (1992) *The Thought of Thomas Aquinas*, Oxford: Clarendon Press, p. 49.

32. DPN ch. 2 (13).

33. In BDT q. 4, a. 2, c.

34. In BDT q. 4, a. 2, c.

35. In BDT q. 4, a. 2, c. See also SCG IV ch. 65.

36. In BDT q. 4, a. 2, c.

37. In BDT q. 4, a. 2, c.

38. In BDT q. 4, a. 2, c.

39. In BDT q. 4, a. 2, c. There are some details and interpretive problems for Aquinas's views on individuation that I can't go into here. For some helpful discussions of these details see: Renard, H. (1943) *The Philosophy of Being*. Milwaukee, WI: Bruce Publishing, pp. 216–8; Bobik, J. (1954) 'Dimensions in the individuation of bodily substances', *Philosophical Studies* 4: pp. 72ff., and Wippel 2000, pp. 362–75.

40. ST Ia. q. 119, a. 1 ad5: 'Sicut Philosophus dicit in I *de Generat.*, quando aliqua materia per se convertitur in ignem, tunc dicitur ignis de novo generari: quando vero aliqua materia convertitur in ignem praeexistentem, dicitur ignis nutriri. Unde si tota materia simul amittat speciem ignis, et alia materia convertatur in ignem, erit alius ignis numero. Si vero, paulatim combusto uno lingo, aliud substituatur, et sic deinceps quousque omnia prima consumantur, semper remanet idem ignis numero: quia semper quod additur, transit in praeexistens.' See also: In DGC I, lec. 16, n. 112 and SCG IV, 81 (12). Cf. In DA II, lec. 9, n. 341 (I am grateful to Robert Pasnau for pointing me to this last text).

41. Similarly, two distinct portions of water that were suddenly pushed together would result in a portion of water numerically distinct from the two original portions of water.

42. QDSC q. un., a. 4, c.: 'Totalitas ... secundum quantitatem ... non potest attribui formis nisi per accidens, in quantum scilicet per accidens dividuntur divisione quantitatis, sicut albedo divisione superficiei. Sed hoc est illarum tantum formarum quae coextenduntur quantitati; quod ex hoc competit aliquibus formis, quia habent materiam similem et in toto et in parte; unde formae quae requirunt

magnam dissimilitudinem in partibus, non habent hujusmodi extensionem et totalitatem, sicut animae, praecipue animalium perfectorum.'

43. QDSC q. un., a. 4, ad19: 'In illis animalibus quae decisa vivunt, est una anima in actu, et multae in potentia; per decisionem autem reducuntur in actum multitudinis, sicut contingit in omnibus formis quae habent extensionem in materia.' See also: SCG II ch. 72; SCG II c. 86; In DA I, lec. 14, n. 208; QDP q. 3, a. 12, ad5 and In Met. VII, lec. 16, n. 1635.

44. Note that the principle need not be so strong for lower animals and plants; presumably there are many types of divisions of such substances that would *not* result in the multiplication of substances of the same kind, e.g. if only the very tip of an earthworm were removed from that earthworm.

    For a more detailed treatment of Aquinas on the individuation of non-living substances, including an attempt to defend his views, see Brown 2001.

45. ST Ia. q. 85, a. 7, ob. 3.

46. ST Ia. q. 85, a. 7, ad3. For some discussion of this passage, see Pasnau 2002, p. 383.

47. See also: SCG II ch. 83 (34); QDA q. un., a. 2, ob. 2 and ad2 and QDA q. un., a. 3, ad12 and ad13. Cf. SCG II ch. 75 (6) and SCG II ch. 81 (7). However, even these latter passages can be harmonized with Aquinas's rejection of 3 in IH.

48. See, e.g.: QDA q. un., a. 1, ad2; ST Ia. q. 76, a. 2, ad 2 and *Responsio ad magistrum Joannem de Vercellis de 108 articulis*, a. 108.

49. ST Ia. q. 76, a. 2, ad2.

50. See, e.g., DEE ch. 6 (93); SCG II ch. 75 (6); QDA q. un., a. 1, ad2 and ST Ia. q. 76, a. 2, ad2.

51. DEE ch. 6 (93).

52. See, e.g., SCG II ch. 83.

53. See, e.g., SCG II chs 80–1 (7).

54. Use of the term 'identity' in this context might be a bit of a stretch, since Aquinas thinks that only substances, strictly speaking, can be identical (see, for example, In Meta. V, lec. 11, n. 907), and artefacts are not substances. Nevertheless, I am going to follow contemporary conventions by using the expression 'identical to' to refer to oneness of artefacts at a time and over time. The reader should simply keep in mind the caveat that artefacts are not substances.

55. I say 'artefact-configuring form' since any substance that functions as the matter of an artefact will presumably have other accidents that have nothing to do with that substance being the matter of an artefact, e.g. extrinsic relations with other objects, or else, accidents that an artefact (or, more properly, the matter of the artefact) can lose without affecting the identity of the artefact in question, e.g. having these determinate dimensions (most would admit that even statues can survive certain variations in shape).

56. In Sent. IV, d. 44, q. 1, a. 1, sol. 2, ad4. See also, In Sent. IV, d. 44, q. 1, a. 1, sol. 3 (ST Suppl., q. 79, a. 3, c.) and QQ 11 q. 6, a. 1, ob. 3 and ad3.

57. See, e.g., In Sent. I, d. 9, q. 1, a. 1. c.; QDP q. 9, a. 1, ad8; In Meta. X, lec. 4, n. 2007; QDSC q. un., a. 3, ad19; ST Ia. q. 29, a. 1, c.; ST Ia. q. 39, a. 3, c.; ST IIaIIae. q. 24, a. 5, ad1 and ST IIIa q. 77, aa. 1 and 2.

58. See, e.g., ST Suppl., q. 79, a. 2, ad4; SCG II ch. 83 (37); SCG IV ch. 80 (3), and In Phys. V, lec. 6, n. 700.
59. See my discussion of the accidental forms of order and composition in Chapter 4.
60. See, e.g., In Sent. IV d. 44, q. 1, a. 1, sol. 2, (end of) ad2.
61. See, e.g., Baker 2000, pp. 36–7.
62. I leave to one side the following complication: Aquinas might very well consider a community or an organization to be artefacts that are composed of human beings.
63. See, e.g.: Burke, M (1980) 'Cohabitation, stuff, and intermittent existence', *Mind* 89: 391–405.
64. DPN ch. 1 (6).

6

Aquinas and the Problem of Material Constitution

In the past three chapters, I have developed an interpretation of some key parts of Aquinas's metaphysic of material objects, including his views on the nature of material substances and artefacts, the different ways in which material substances and artefacts are composed, and the identity and individuation conditions for material substances and artefacts. With these chapters providing the necessary metaphysical background, I now turn to developing Thomistic solutions to the classical and contemporary puzzles about material objects that raise the PMC.

The Ship of Theseus puzzle

In Chapter 1, I suggested that puzzles about compound material objects such as the Ship of Theseus appear to raise the following problem: our common-sense intuitions about compound material objects are logically inconsistent with one another. This apparent conflict between common sense and logic is known as 'the problem of material constitution' (PMC). The puzzles about material objects that raise the PMC all assume that the following common-sense intuitions about material objects are true:

(IMO1) There are such things as *compound material objects*, that is there are material objects that are composed or constituted of other material objects.

(IMO2) There are many, many different kinds of compound material objects, including different kinds of atoms, molecules, aggregates of atoms or molecules, proteins, enzymes, plants, animals, tissues, organs, limbs, body sections, artefacts and artefact parts.

(IMO3) Compound material objects endure through time and change.

(IMO4) There are compound material objects that can survive certain losses, gains and replacements with respect to their parts.

(IMO5) Two material objects cannot exist in the same place at the same time.

(IMO6) Necessarily, identity is a transitive relation.

In offering my proposed Thomistic solutions to those puzzles about material objects that raise the PMC, I begin with the most famous: the Ship of Theseus puzzle.

Recall that the Ship of Theseus is a sailing ship composed of wooden planks. After a few years, the ship's planks begin to weather. The crew in charge of the Ship of Theseus decides to lay down a policy: before the Ship of Theseus sets sail each year, the weathered planks of the ship will be replaced with new ones. After a few years, all of the planks of the original ship are replaced. However, someone (let's say her name is Merry) collects the planks disposed from the original ship each year, until eventually Merry has collected all of the planks from the original ship. Furthermore, Merry decides to put the planks (from the original ship) together in her back yard, giving those planks the same sort of distinctive configuration they had when they composed the original ship at the time of her first voyages. Which ship is identical to the original ship? Is it the *continuous ship*, whose spatio-temporal history is continuous with that of the original ship, or is it the *reconstructed ship*, which is composed of the same set of planks as the original ship?

Given the common-sense intuitions about material objects I enumerated above, the Ship of Theseus puzzle can be formulated in such a way that it explicitly raises the PMC, and I call this formulation 'SOT'. SOT employs the following nomenclature. At time t, we have 'the original ship'. Now the original ship is composed of an aggregate of planks at t. Call this aggregate, 'aggregate$_{OS}$'. At time $t + 1$, there is 'the continuous ship' and 'the reconstructed ship'. These ships are also each composed of an aggregate of planks (call them 'aggregate$_{CS}$' and 'aggregate$_{RS}$', respectively). SOT thus runs as follows.

(1) The original ship at t is numerically identical to aggregate$_{OS}$ [from IMO3 and IMO5].

(2) Aggregate$_{OS}$ is numerically identical to aggregate$_{RS}$ [from the assumption that aggregates x and y are numerically identical if, and only if, x and y have all and only the same proper parts].

(3) Aggregate$_{RS}$ is numerically identical to the reconstructed ship at $t + 1$ [from IMO3 and IMO5].

(4) Therefore, the original ship at t is numerically identical to the reconstructed ship at $t + 1$ [from (1)–(3) and IMO6].

(5) The original ship at t is numerically identical to the continuous ship at $t + 1$ [from the fact that the original ship at t and the continuous ship at $t + 1$ are spatio-temporally continuous and common-sense intuitions IMO2, IMO3 and IMO4].

(6) Therefore, the reconstructed ship at $t + 1$ is numerically identical to the continuous ship at $t + 1$ [from (4), (5) and IMO6].

(7)   The reconstructed ship at $t+1$ is not numerically identical to the continuous ship at $t+1$ [from the impossibility of bi-location].

Now, (6) and (7) are contradictory opposites and (6) follows logically from (1)–(5). But (7) looks undeniably true – it is impossible for a material object to bi-locate – whereas premises (1)–(5) are all entailments of common-sense intuitions about material objects. SOT suggests that our common-sense conception of a compound material object is paradoxical.

How would Aquinas respond to SOT? Aquinas's metaphysic of material objects gives him at least two possible strategies for solving the PMC in so far as it is raised by SOT. I will call these 'the reductivist strategy' and 'the non-reductivist strategy', respectively.

One way of handling the paradox arising from the premises of SOT is to deny that there are any such things as ships and, for that matter, ship-planks, because there simply are no such things as artefacts. Of course, this claim need not be construed as identical to the (absurd) claim that there is *nothing* existing where what we normally refer to as a 'ship' is located. Instead, the philosopher taking up the position that there are no ships can say that there is never any single material object where what we refer to as a 'ship' is located. Instead, where what we call a 'ship' is located there is just a plurality of material objects (where 'plurality' should not be taken – despite its grammatical form – to signify an individual thing).

As we have seen, Aquinas takes up a similar sort of position on the ontological status of artefacts. Although he does not deny them being *simpliciter*, he does deny that artefacts are substances. For Aquinas, an artefact simply is a composite of a material substance (or, in the case of the Ship of Theseus, a number of material substances) and a particular accidental form. The name we give to an artefact, e.g. 'Ship of Theseus', in reality refers to a plurality of material substances *insofar as they are accidentally modified in a certain way*. We might say that Aquinas explains the material existence of an artefact *reductively*, that is, in terms of the substance or substances that compose it.

According to the reductivist strategy, where SOT goes awry is in premise (5). That premise assumes that artefacts are objects that Aquinas would refer to as 'substances': material objects that can survive (all of) the replacement of their integral parts. But the original ship is not numerically identical to the continuous ship for Aquinas, since the original ship's matter is not identical to the matter of the continuous ship. This is because the matter of the original ship is completely different from the matter of the continuous ship: the aggregate of substances that makes up the matter of the original ship has no members in common with the aggregate of substances that makes up the matter of the continuous ship.[1]

Because the matter of the original ship and the matter of the continuous ship are completely different, and because, as we saw in Chapter 5, the matter of an artefact individuates an artefact's form, it follows that the accidental forms of the original ship and the continuous ship are not the same. But identity of form is a necessary condition for artefact identity according to Aquinas.[2] Therefore, he thinks that the original ship is not numerically identical to the continuous ship. The Ship of Theseus's paradoxical result – that the Ship of Theseus can be bi-located – is thereby undermined by the reductivist strategy. And it is undermined on the basis that an artefact has an accidental form as its form of the whole; artefacts are not substances for Aquinas.

Some may think that the view that the original ship is not identical to the continuous ship – that an artefact is not the sort of thing that can survive replacement of its parts, i.e. a substance – is rather counter-intuitive. Whether or not this is so – and I'll have more to say on this score in the sequel – Aquinas's metaphysical commitments allow him to take up an alternative strategy to solving puzzles such as the Ship of Theseus, one that I will call the 'non-reductivist strategy'. As the name indicates, one advantage of this approach to the Ship of Theseus puzzle is that it allows Aquinas to solve the puzzle while admitting, if only for the sake of argument, that artefacts enjoy the same ontological status as trees and tigers.

The non-reductivist strategy challenges the truth of premise (2) of SOT. There are two ways of interpreting premise (2) of SOT. One of these involves construing the term 'aggregate' in that premise as referring to an object y composed of parts (call them 'the xs') where y exists *no matter what* the spatial relations that obtain among the xs. In that case, an aggregate would be an individual object that exists simply because its parts exist. According to this understanding of 'aggregate', pick any two objects in the universe, e.g. the sun and my left shoe, and they form a third individual object, that is, an aggregate-object. Call this the *universalist* interpretation of 'aggregate'.[3] That there are such material objects is highly counter-intuitive. The ontology of someone that accepts that any two material objects make for a third material object would be excessively bloated, to say the least.

A second way to interpret premise (2) of SOT involves understanding the term 'aggregate' to refer to any material object y that is composed of the xs, where the xs are such that they enter into certain kinds of spatial and/or causal relations to one another. For example, perhaps the xs form an aggregate-object y whenever the xs are sufficiently bonded to one another and each of the xs is still the same being it was prior to being bonded – take a bundle of sticks as a prime example of an aggregate in this sense. Call this way of understanding 'aggregate' in premise (2) of SOT the *bonding* interpretation.

But if we understand premise (2) of SOT in accord with the bonding interpretation of 'aggregate', then (2) is true only if it is possible for objects to exist intermittently. Recall from Chapter 5 that the Intermittent Existence Thesis has it that:

(IE) It is possible that x exists at t, x goes out of existence at $t + 1$, y begins existing at $t + 2$ and y is numerically identical to x.

Call the original planks of the Ship of Thesus 'the xs'. Given the bonding interpretation of 'aggregate', premise 2 entails the following: (a) the xs compose an aggregate y at t (aggregate$_{OS}$), (b) the xs compose an aggregate z at $t + 1$ (aggregate$_{RS}$), (c) y is identical to z and (d) the xs do not compose a material object between t and $t + 1$ (for some of the xs are out to sea, some of the xs are in Merry's back yard, and according to the bonding interpretation of 'aggregate', the xs must be bonded in order to compose an aggregate). Premise (2) of SOT entails that objects can intermittently exist.

And, as I have suggested in Chapter 5, Aquinas certainly rejects that intermittent existence is possible. Since the most plausible understanding of premise (2) of SOT assumes that objects can intermittently exist, Aquinas would reject that premise.

Furthermore, although I can't give a thorough defence of the falsity of IE here, it seems to me that IE is just as counter-intuitive as thinking that it is possible for two material objects to coincide spatially.[4] Thus, one accepts that premise 2 of SOT is true – on the bonding interpretation – only by admitting a counter-intuitive thesis about material objects.

But need we accept the truth of premise (2) of SOT? The reader will note that premise (2) of SOT is not entailed by any common-sense intuitions about material objects that I outlined above (IMO1–IMO6). No other common-sense intuitions about material objects entail (2). Also, we can better state the identity conditions for aggregates by noting that aggregates x and y being numerically identical have sameness of parts (members) as a necessary (but not a sufficient) condition. So rejecting premise (2) does not conflict with our common-sense intuitions about material objects (including aggregate-objects).

Note that premise (2) of SOT commits one to holding a counter-intuitive view, no matter how the term 'aggregate' is read in that premise. According to the universalist understanding of premise (2) of SOT, any two (or more) material objects compose a material object. But that entails that the sun and my left shoe compose a distinct material object. But common sense has it that there is no distinct material object composed just of the sun and my left shoe. Therefore, common sense has it that universalism is false. Clearly the universalist understanding of premise (2) of SOT commits one to a counter-intuitive view of material objects.

According to the bonding interpretation of premise (2) of SOT, premise (2) entails IE. But common sense has it that IE is false – it's simply not possible for objects to go out of existence and come back into existence. To say that material objects can intermittently exist is to say something that is in conflict with our ordinary ways of thinking about material objects. Therefore, the bonding interpretation of premise (2) of SOT – like the universialist interpretation – conflicts with common sense.

But even if IE itself is not counter-intuitive as I have been suggesting, certainly it must be admitted that the *denial* of IE – saying it is impossible for objects to exist intermittently – is not in conflict with common sense. Therefore, rejecting premise (2) of SOT on the grounds that it commits one to the truth of IE certainly counts as a solution to SOT that does not depend upon rejecting any of our common-sense intuitions about material objects. There is thus good reason – it is in accord with our common-sense intuitions about material objects – to solve the Ship of Theseus puzzle by rejecting premise (2) of SOT.

Aquinas would therefore reject premise (2) of SOT on the grounds that it commits one to a counter-intuitive view about material objects. Thus, in addition to the reductivist strategy for solving SOT, Aquinas can solve SOT without assuming that artefacts are not substances by way of the non-reductivist strategy.

However, the non-reductivist strategy cannot be used to solve all of the classical and contemporary puzzles that philosophers talk about as raising the PMC. Indeed, a variant on SOT can be formulated that raises the PMC afresh (call it 'SOT*'). SOT* has premises that do not allow one to solve it by way of the non-reductivist strategy (note that some of SOT*'s premises are taken from SOT):

(1)   The original ship at t is numerically identical to aggregate$_{OS}$ [from IMO3 and IMO5].

(5)   The original ship at t is numerically identical to the continuous ship at $t+1$ [from the fact that the original ship at t and the continuous ship at $t+1$ are spatio-temporally continuous and common-sense intuitions IMO2, IMO3 and IMO4].

(8)   The continuous ship at $t+1$ is numerically identical to aggregate$_{CS}$ [from IMO3 and IMO5].

(9)   Aggregate$_{OS}$ is not numerically identical to aggregate$_{CS}$ [from the fact that aggregate$_{OS}$ and aggregate$_{CS}$ have different sets of proper parts and an accepted identity condition for aggregates: aggregates x and y are numerically identical only if x and y have all and only the same proper parts].

(10)   The original ship at t is not numerically identical to the continuous ship at $t+1$ [from (1), (8), (9) and IMO6].

Premises (5) and (10) are contradictory opposites and seem to follow from premise (9) and our common-sense intuitions about material objects. But premise

(9) of SOT* can't be rejected for the sort of reason that might lead one to reject premise (2) of SOT. Unlike premise (2) of SOT, premise (9) of SOT* looks *undeniably* true. This is because all (9) says is that aggregate$_{OS}$ is *not* numerically identical to aggregate$_{CS}$, and this because a necessary (although not sufficient) condition for aggregates x and y being numerically identical is that x and y have the same proper parts (members). But aggregate$_{OS}$ and aggregate$_{CS}$ clearly do not have the same proper parts. Thus, SOT* raises the PMC in a way that does not allow one to employ the non-reductivist strategy.

Where does this leave the non-reductivist strategy for solving SOT? First, let me point out that some of the assumptions at play in SOT* are not typically the assumptions that are the focus of the Ship of Theseus puzzle. Indeed, SOT* is really just a version of *another* puzzle about material objects known as the Debtor's Paradox. Indeed, I freely admit that the non-reductivist strategy for solving the Ship of Theseus puzzle has nothing to say as far as solving the Debtor's Paradox is concerned. Although the Ship of Theseus puzzle and the Debtor's Paradox both raise the PMC, they clearly do so in different ways and that is why philosophers working on problems of material constitution typically treat them as two *different* puzzles about material constitution.[5] Thus, the inability of the non-reductivist strategy to solve SOT* (or the Debtor's Paradox, or the puzzle of Tibbles the Cat) does not render it impotent as a solution to the Ship of Theseus puzzle.

I can make this defence of the non-reductivist strategy in another way. To invoke some terminology coined by Michael Rea, the non-reductivist strategy constitutes only a *partial* solution to the PMC. According to Rea's way of categorizing solutions to the PMC, a proposed solution to the problem is *complete* if it can solve every puzzle about material objects that raises the PMC. In contrast to a complete solution to the PMC, a *partial* solution to the PMC can solve only one or some of the puzzles that raise the PMC.[6] I admit that the non-reductivist strategy offers only a partial solution to the PMC. Nonetheless, the non-reductivist strategy does allow one to solve the Ship of Theseus puzzle in a way that is fully in accord with our common-sense intuitions about compound material objects, and it therefore constitutes a powerful solution to that problem.

The Debtor's Paradox

Since SOT* is really a version of the classical puzzle about material objects known as the Debtor's Paradox, let's turn to that puzzle next. Like the Ship of Theseus puzzle, the Debtor's Paradox raises questions about the identity relationships that hold between non-aggregate material objects and aggregates of material objects. However, in contrast to the Ship of Theseus's focus on artefacts, the Debtor's Paradox is a puzzle having to do with *organisms* and their parts.

Here's how the puzzle goes. Imagine that Chris borrows some money from Merry at time t. At t, Chris is constituted of an aggregate of fundamental particles, call it 'aggregate$_{\text{at } t}$'. At a later time, $t+1$, Chris is also constituted of an aggregate of fundamental particles, call it 'aggregate$_{\text{at } t+1}$'. But Chris has replaced many of his physical parts between t and $t+1$, and so aggregate$_{\text{at } t}$ is not identical to aggregate$_{\text{at } t+1}$. At $t+1$, Merry comes to collect her debt from Chris (she assumes that Chris at t is numerically identical to Chris at $t+1$). Chris argues that he couldn't be the person who borrowed money from Merry, since he is identical to an aggregate of fundamental particles, the person who borrowed money from Merry was constituted of a different set of fundamental particles than he, and aggregates that have different parts as members cannot be identical. Can Merry believe that Chris at t is identical to Chris at $t+1$ without denying an obvious truth about aggregate-identity? More formally, the Debtor's Paradox can be shown to raise the PMC as follows (call this formulation of the Debtor's Paradox, 'DP'):

(1)   Aggregate$_{\text{at } t+1}$ is numerically identical to Chris at $t+1$ [from IMO3 and IMO5].

(2)   Chris at $t+1$ is numerically identical to Chris at t [from the fact that Chris at $t+1$ is spatio-temporally continuous with Chris at t and common-sense intuitions IMO2, IMO3 and IMO4].

(3)   Chris at t is numerically identical to aggregate$_{\text{at } t}$ [from IMO3 and IMO5].

(4)   Therefore, aggregate$_{\text{at } t}$ is numerically identical to aggregate$_{\text{at } t+1}$ [from 1–3 and IMO6].

(5)   But aggregate$_{\text{at } t}$ is not numerically identical to aggregate$_{\text{at } t+1}$ [from the fact that aggregate$_{\text{at } t}$ and aggregate$_{\text{at } t+1}$ do not share all and only the same proper parts and a standard assumption about the identity criteria for aggregates: aggregates x and y are numerically identical only if x and y have all and only the same proper parts].

Aquinas would have a number of things to say about DP. But first what Aquinas would not say. Of course, he wouldn't deny (2). Thus, Aquinas could not employ the reductivist strategy where DP is concerned, since Chris is a human organism, and human organisms can survive part replacement according to Aquinas – even a total replacement of their parts. This is because Aquinas thinks that human organisms x and y are numerically identical if (and only if) x and y have the same substantial form. Thus, x and y could have completely different sets of integral parts for Aquinas. Finally, I see no reason why Aquinas wouldn't accept premise (5) of DP. Aggregates are like sets in that their identity is tied to their component parts, or members. If aggregates x and *y* are numerically identical, then *x* and *y* have all and only the same proper parts.

The premises of DP that Aquinas would reject are (1) and (3). A substance *x* is not identical to the aggregate of integral parts that composes *x* for Aquinas. This is because substances such as human beings have a substantial form that preserves their identity through (at least many sorts of) changes with respect to their integral parts. Organisms are objects that can change out their parts, whereas aggregates cannot lose or gain a part without going out of existence (at least on the most plausible account of aggregate identity). Since Chris and Merry are essentially organisms, neither Chris nor Merry could be identical to an aggregate-object according to Aquinas.[7]

Although Aquinas would deny premises (1) and (3), he could do so without denying IMO3 and IMO5. (If this is right, then there may be more leading someone to make the identity claims in (1) and (3) than fear of temporal parts or spatially co-incident objects). Aquinas can deny that there is an aggregate-object existing wherever a substance exists, and do so without denying that aggregate-objects exist *simpliciter*.

An aggregate is a collection of objects, where each member (part) of the aggregate is at least every bit as much a real object as the aggregate itself. To put it in Aquinas's terms, an aggregate is a collection of *substances* modified by a form of the whole that is an accidental form. On Aquinas's account of substance, every substance has only one substantial form, and this substantial form confers existence and specific identity on the substance itself as well as on all of its integral parts. Since the identity of an integral part depends upon the existence of the whole to which it belongs, the integral parts of a substance are not themselves substances. Now, because the integral parts of a substance *are not* substances and the integral parts of an aggregate *are* substances, there simply is no aggregate-object existing in the same place at the same time as any organism. Say there is an aggregate *a* composed of non-living substances at *t* (call the substances that compose *a* 'the *x*s'). If a substance *y* is generated out of the *x*s at *t* + 1, then the *x*s no longer exist at *t* + 1, and so neither does *a*. Aquinas thus solves DP (and SOT*) – not by denying that aggregates exist *simpliciter*, but by giving good reasons for denying that there is an aggregate-object existing wherever a substance (such as a living organism) exists.[8]

But this is not the only way of denying the identity claims in (1) and (3) without thereby admitting spatially coincident objects.[9] As we have seen in Chapter 2, Peter van Inwagen has reasons for believing there simply are no aggregate-objects. Furthermore, Van Inwagen takes it that a compound material object is related to the parts that compose it in a one–many relation, which he considers to be an innocent kind of spatial coincidence. Although Van Inwagen's views are similar to Aquinas's – both reject that an organism is composed of an aggregate-object that is non-identical to that organism – Van Inwagen's views have a weakness that do not affect those of Aquinas.

To see why, consider the following puzzle, which is a variation on DP:

> Joe is a living human being and is composed of some cells at time t (call that precise group of cells that compose Joe at t, 'the xs'). Joe dies an instant later at $t+1$ and Joe's corpse at $t+1$ is composed of some cells (call these 'the ys'). Since Joe at t and Joe's corpse at $t+1$ are spatio-temporally continuous, the xs are identical to the ys.

Given the claims in the above story, one of the following two statements is true:

> (a) Since the ys compose an individual material object at $t+1$ (call it 'the corpse's aggregate'), the xs therefore compose an individual material object at t (call it 'Joes's aggregate') and Joe's aggregate is not identical to Joe at t (since organisms and aggregates have different essential properties) and Joe's aggregate is spatially coincident with Joe at t.

> (b) Neither the xs nor the ys compose an *individual* material object, but the xs are nonetheless spatially coincident with an individual material object, Joe at t.

Thus, whether one accepts (a) or (b), one is accepting the possibility of *some* sort of spatial coincidence. Call this puzzle – which is specifically aimed at the philosopher who wants to *avoid* admitting the possibility of spatially coincident material objects – 'the puzzle of the Dead Debtor', or 'DD' for short.

Before explaining how Van Inwagen and Aquinas would respond to DD, I want to make mention of a possible response to DD that Frederick Doepke calls the 'dichronic view'.[10] Although the ys are temporally and spatially continuous with the xs, the ys are not identical to the xs according to the dichronic view. As we shall see, Aquinas's solution to DD shares this particular feature with the solution of the dichronic view. But what explains the continuity of the ys and the xs according to the dichronic view? Apparently nothing explains it; the continuity of the ys and the xs is taken to be a brute fact confirmed by observation according to the dichronicist. The dichronic view rejects the assumption that there needs to be a substratum for any change. But this solution to DD suffers the problem of sacrificing the intelligibility of substantial change. As I argued in Chapter 4, the intelligibility of any change requires a substratum for that change, including radical sorts of change such as the substantial changes that occur when the xs go out of existence and the ys come into existence. Thus, the dichronicist's solution to DD suffers the problem of sacrificing the intelligibility of the changes that occur in DD.

In contrast to the dichronicist, Van Inwagen accepts (the possibility that) the *x*s are identical to the *y*s. How would Van Inwagen respond to DD? Van Inwagen rejects that aggregates are in any sense individual material objects. He thus rejects possibility (a) above. But, given (b), which Van Inwagen does accept, it is the case that the *x*s are spatially co-incident with Joe at *t*. Thus, (what I imagine would be) Van Inwagen's response to DD still leaves a one–many coincidence. Perhaps one will think this type of co-incidence innocuous. Indeed, one–many co-incidence is certainly not objectionable for the same reason that two individual objects existing in the same place at the same time is objectionable (one–one co-incidence).[11] But what one–many co-incidence does seem to undermine is the unity enjoyed by the organism that is coincident with the *x*s. If the *x*s exist as (the equivalent of) material substances when they compose Joe, then Joe looks to be merely a derivative identity, that is a property of a plurality of material objects, and not a material object himself. Thus, although taking option (b) in understanding the case of the Dead Debtor may absolve one from the objections aimed at one–one spatial co-incidence, it nevertheless puts into doubt the substantial ontological status of composed material objects.

Aquinas's understanding of material composition does not suffer this weakness: the integral parts of substances are only potential and not actual substances as long as they remain parts of a substance. Aquinas's views thus preserve the view that Joe is composed of cells at *t*, but it also preserves Joe's ontological status as a substance at *t*. Joe is not a derivative entity. In fact, things are the other way around according to Aquinas: it is the cells that compose Joe that are the derivative entities, and not human beings such as Joe.

The Thomistic solution to DD preserves the ontological status of Joe as a full-blown substance while also making sense of substantial change. Like the dichronicist, Aquinas would reject the assumption on the part of DD that it is possible that the *x*s are identical to the *y*s. For Aquinas, this is because each one of the *x*s is a part of a substance, whereas (assuming a corpse is an aggregate of substances) each one of the *y*s is an actual substance. But a part of a substance is not identical to a substance. Therefore, none of the *x*s are identical to any of the *y*s. However, Aquinas does not think these changes come about without a substratum, as the dichronicist assumes. As we saw in Chapter 4, prime matter is the subject for such radical changes as cells being transformed into numerically different cells (which in this case is a function of the corruption of a human being such as Joe). There is a substratum for the change in DD according to Aquinas, but that substratum is not a substance or an aggregate of substances, but matter without any form.

Thus, Aquinas's metaphysics (specifically, his doctrine of metaphysical parts) allows him to solve DD without admitting coincident objects, denying that Joe is a compound material object, denying Joe's ontological status as a substance (a unified being in the strongest sense), or denying that change always requires a substratum. Nor does Aquinas's solution to DD require admitting an eccentric understanding of the identity relation or the existence of temporal parts. Aquinas's metaphysics allows him to solve puzzles about compound material objects while preserving common-sense intuitions about such objects.

The Growing Argument

There is another classical puzzle having to do with compound material objects that is a close cousin of the Debtor's Paradox. This puzzle is known as the Growing Argument. Like the Debtor's Paradox and the Ship of Theseus, the Growing Argument is a puzzle about material objects that goes back to antiquity (hence I have been speaking about 'classical' puzzles about material objects).[12] The Growing Argument specifically tries to show that a material object cannot survive the *addition* of a proper part. To see why, say that John is a growing boy, and John gets bigger between time t and $t + 1$. Also, say that 'a' picks out John's improper part at time t and '$a1$' picks out the proper part of John at time $t + 1$ that fills precisely the same space as a did at t. Employing these variables, the Growing Argument can be formulated in such a way that it explicitly raises the PMC (call this version of the Growing Argument, 'GA'):

(1) John at $t + 1$ is numerically identical to John at t [from the fact that John at $t + 1$ and John at t are spatio-temporally continuous and common-sense intuitions IMO2, IMO3 and IMO4].

(2) John at t is numerically identical to a [from the definition of 'improper part' – x's improper part is numerically identical to x – and common-sense intuitions IMO2 and IMO5].

(3) a is numerically identical to $a1$ [from the fact that a and $a1$ are spatio-temporally continuous and IMO2].

(4) John at $t + 1$ is numerically identical to $a1$ [from (1)–(3) and IMO6].

(5) John at $t + 1$ is not numerically identical to $a1$ [from the fact that $a1$ is a proper part of John at $t + 1$ and the definition of 'proper part': a proper part of x is a part of x that is not numerically identical to x].

Premise (3) of GA looks true if we assume that there are objects such as $a1$ (IMO2). For object a does not suffer any intrinsic changes when John gains a part at $t + 1$. Since it is also the case that $a1$ fills the same space at $t + 1$ that a fills at t, it is reasonable to assume that the relation between a and $a1$ is that of identity. Premise (2) assumes that coincident objects are not to be countenanced (IMO5) and that there are material objects such as John. Premise (1) is based on the by now familiar line that organisms such as John endure through changes – even changes with respect to their proper parts (IMO3 and IMO4). Finally, premise (4) follows from premises (1)–(3) and the assumption that identity is necessarily transitive (IMO6). But premise (5) – the contradictory opposite of (4) – is a necessary truth. It simply follows from the uncontroversial definition of a proper part. Proper parts have that feature normally associated with integral parts, that is, proper parts are smaller than the wholes to which they belong. A proper part is a part of x that is not identical to x. Hence, GA generates the PMC.

Aquinas is, of course, committed to the possibility of (1). Furthermore, assume that he would accept the contemporary mereological notions of 'improper part' and 'proper part'. Aquinas must reject premise (3) of GA. And indeed, he would, but not (necessarily) for the reasons given by some contemporary philosophers. For example, Van Inwagen rejects (3) because it wrongly assumes that there are material objects such as $a1$ – $a1$ being the proper part of Joe at $t+1$ that fills precisely the same space as a, where a is Joe's improper part at t. From Van Inwagen's perspective, although there certainly is a plurality of material objects existing in the space supposedly occupied by $a1$, there is no individual object in that space. According to Van Inwagen, $a1$, if it existed, would be what he calls 'an arbitrary undetached part', and he has argued that we should not believe in the existence of such parts. Indeed, he takes arguments such as GA to be a *reductio ad absurdum* of the belief in arbitrary undetached parts.[13] As I noted in Chapter 4, it is not entirely clear to me whether Aquinas accepts the existence of arbitrary undetached parts. If he does not, then Aquinas could answer GA in the same manner as Van Inwagen.

But Aquinas need not take up this eliminativist line to solve GA, for he has something else to say about why (3) is false. Contrary to what the puzzle assumes, a and $a1$ are not identical, since a is a substance and $a1$ is not a substance. At most, $a1$ is only an integral part of a substance, and no integral part of a substance is identical to a substance for Aquinas. How does Aquinas then explain the spatial and temporal continuity of a and $a1$? Aquinas's view here can be defended if we recall his take on the relationship between Joe and Joe's corpse in the analysis of argument DD. Joe and Joe's corpse are also spatially and temporally continuous, but there is no sense in which Joe is identical to a corpse. That is it is possible that Joe exists in space s at t, and Joe's corpse exists in s at $t+1$ and Joe is not identical to Joe's corpse. And this denial that Joe is not identical to Joe's corpse makes perfect sense. Analogously, it makes sense to say with Aquinas that a exists in space s at t and $a1$ exists in s at $t+1$ and a is not identical with $a1$. The advantage of this solution to GA over Van Inwagen's eliminativist solution is that it preserves an additional common-sense intuition about material objects, namely that there are such things as arbitrary undetached parts.

The Puzzle of Tibbles the Cat

The next puzzle I want to look at is a contemporary one, but it is a close relative of the Growing Argument. This puzzle switches the emphasis from questioning the purported ability of a material object to *gain* a part to questioning the purported ability of a material object to survive the *loss* of a part. The puzzle in question goes by various titles, but I will refer to it as 'the Puzzle of Tibbles the Cat'.[14] Tibbles is a cat, and so is a compound material object. One of the parts that Tibbles has at time *t* is her tail. Just as Tibbles's tail is a proper part of Tibbles at *t*, so is that part of Tibbles at *t* that includes all of Tibbles except her tail (assume that it is perfectly clear where Tibbles's tail ends and begins). Call the part of Tibbles minus her tail, 'Tib'. Now imagine that at time *t* + 1, Tibbles does not have a tail, having lost it in the process of trying to escape from an on-coming car sometime between *t* and *t* + 1. With this much information, the PMC can be generated as follows (call the following formulation of the Tibbles puzzle that explicitly generates the PMC, 'TC'):

(1)   Tibbles at *t* is numerically identical to Tibbles at *t* + 1 [from the fact that Tibbles at *t* and Tibbles at *t* + 1 are spatio-temporally continuous and common-sense intuitions IMO2, IMO3 and IMO4 – a cat can survive the loss of her tail].

(2)   Tibbles at *t* + 1 is numerically identical to Tib at *t* + 1 [from IMO2 and IMO5].

(3)   Tib at *t* + 1 is numerically identical to Tib at *t* [from the fact that Tib does not undergo any intrinsic changes when an object extrinsic to Tib – Tibbles's tail – changes its spatial location and common-sense intuition IMO2].

(4)   Therefore, Tibbles at *t* is numerically identical to Tib at *t* [from (1)–(3) and IMO6].

(5)   But Tibbles at *t* is *not* numerically identical to Tib at *t* [since Tib at *t* is a proper part of Tibbles at *t*, and by definition no proper part of *x* is numerically identical to *x*].

TC purports to show that an organism that changes its parts is both identical and non-identical to one of its proper parts at a certain time. Since this is impossible, one of TC's premises must be false. But all of TC's premises appear to be entailed by common-sense beliefs about compound material objects. Our common-sense beliefs about material objects seem to entail a contradiction. TC casts doubt on the intelligibility of the common-sense concept of a compound material object.

Aquinas may solve TC in the same way that Van Inwagen does, by denying the existence of objects such as Tib. Tib is an arbitrary undetached part of

Tibbles, a part of Tibbles analogous to the north side of my house. But there is no single material object that is the referent of 'the north side of my house'. Thus, the same is true for the referent of 'Tib'. Therefore, those premises of TC – e.g. (2) and (3) – that assume that Tib is an individual material object are to be counted either as false or ill-formed.

But there certainly are some objects that Van Inwagen styles as arbitrary undetached parts of which Aquinas would not deny the existence, namely *functional* integral parts such as hands, brains and heads.[15] Even if hands, brains and heads are not substances, they do nonetheless have being in an extended sense according to Aquinas. Nonetheless, if a Thomist is tempted to solve TC in eliminativist fashion by denying the existence of arbitrary undetached parts such as Tib, a puzzle analogous to TC can be constructed that would not allow (the Thomist) such a strategy. Consider the following case:

> Jane is a typical human being at time t, but (literally) loses her head in a car accident at $t+1$. Through an incredible sequence of events (this is admittedly a science-fiction sort of example), Jane's decapitated head is delivered alive to a hospital and hooked up to a machine that keeps it alive at $t+2$. Imagine that Jane is thus alive and well at $t+2$.

We might thus construct a puzzle analogous to the Tibbles puzzle called 'the puzzle about Jane's head ('JH' for short):

(1)   Jane at t is numerically identical to Jane at $t+2$ [from the fact that Jane at t and Jane at $t+2$ are spatio-temporally continuous and common-sense intuitions IMO2, IMO3 and IMO4 – and assuming that a human being can survive the loss of all her integral parts save her head].

(2)   Jane at $t+2$ is numerically identical to Jane's head at $t+2$ [from IMO2 and IMO5].

(3)   Jane's head at $t+2$ is numerically identical to Jane's head at t [assuming that an object is not changed when an object extrinsic to it – in this case, the rest of Jane's body – changes its spatial location and common-sense intuition IMO2].

(4)   Therefore, Jane at t is numerically identical to Jane's head at t [from 1–3 and IMO6].

(5)   But Jane at t is not numerically identical to Jane's head at t [from the fact that Jane's head at t is a proper part of Jane at t and the definition of 'proper part'].

Aquinas would have a lot of things to say about JH. But even if Aquinas were to agree with Van Inwagen that there are no such objects as Tib, Aquinas certainly thinks that there are such things as heads, hands and hearts (contra Van Inwagen). So Aquinas cannot solve JH by denying that such objects exist. However, Aquinas's metaphysic of material objects allows him an alternative solution to JH: premise (3) of JH is false because Jane's head at t is only potentially a substance – or, alternatively, an integral part of a substance – since it is an integral part of Jane at t. But Jane's head at $t+1$ is an actual substance; this in virtue of the fact that Jane now occupies just that space that was once occupied only by her head. Since an integral part is not a substance and a non-substance is not identical to a substance, Jane's head at t is not identical to Jane's head at $t+2$. Aquinas can thus solve TC in the same way: if Tib exists at t, then Tib at t will not be identical to Tib at $t+1$, since Tib at t is not a substance (Tib at t is an integral part of a substance) whereas Tib at $t+1$ is a substance. To put things differently, Tib no longer exists at $t+1$ since Tib is by definition an integral part of a substance, and there is no integral part of a substance (of the type Tib is) at $t+1$.

However, Aquinas has perhaps another solution to JH open to him as well. He can deny premise (2) of JH – that Jane at $t+2$ is identical with Jane's head at $t+2$ – on the grounds that saying that Jane at $t+2$ is identical to Jane's head at $t+2$ entails that Jane was *all along* identical only to her head – that Jane's torso, arms and legs were never *really* parts of her, but only objects physically contiguous to Jane. I think Aquinas would agree with Van Inwagen that what should rather be said about Jane's identity is that Jane suddenly becomes a lot smaller at $t+2$.[16] Although Jane's head can preserve Jane's identity and existence, Jane can't be identical to her head. In other words, a human being such as Jane can't be identical to her head, although she may be composed of only her head. As I have been suggesting throughout, Aquinas accepts the metaphysical thesis that says 'composition is not identity'.

It can be argued that Aquinas explicitly thinks this way about a case that is analogous to that of Jane and her head. Consider a case where, instead of surviving the car accident, Jane dies as a result of it. According to Aquinas, Jane does not cease to exist at death. Jane's soul, which is a metaphysical part of her, continues to exist after Jane's death and preserves Jane's existence and identity. However, Aquinas thinks that Jane is not identical to her soul. Therefore, Aquinas believes that Jane's soul is sufficient to preserve Jane's identity through time and change – Jane even continues to exist after her biological death for Aquinas – but that Jane is not (indeed, Jane can't be) numerically identical to her soul since Jane is essentially a rational animal. Thus, Aquinas also accepts the following extension of the thesis that composition is not identity: Jane is not identical to her soul but she can be composed of only her soul.

Notice too that Aquinas's adherence to the principle that composition is not identity allows Aquinas to deny premise (2) of JH *without* thereby committing him to the possibility of spatially co-incident objects. Indeed, someone might reject premise (2) of JH because they think that two individual material objects can exist in the same place at the same time; in this case, Jane's head *and* Jane. This would, of course, get one around the paradox, but at the (high) price of admitting the possibility of spatially co-incident objects. But Aquinas's view that a part of a substance x can sustain the identity of x without being identical to x allows him to deny premise (2) of JH without admitting co-incident objects. At time $t + 2$ there is only one object present in the story of Jane and her head: Jane's head. Nevertheless, Jane's head – although not identical to a human being – is (I am assuming) a part of a human being sufficient to preserve the identity and existence of a human being.

Of course, Aquinas need not employ this kind of solution in the case of Tibbles the Cat. In that case Aquinas could reasonably admit that Tibbles without her tail is numerically identical to Tibbles, since all that has happened to Tibbles at this point is that she has lost her tail. In such a case, Aquinas can remain content *denying* that an integral part of a substance (Tib at t) is numerically identical to a substance (Tib at $t + 1$). In actual fact Tib no longer exists at $t + 1$ since Tib is a part of a substance and what exists at $t + 1$ is an actual substance, namely Tibbles. Although there may indeed be parts such as Tib (arbitrary undetached parts), Tib does not exist at $t + 1$, and so premise (3) of TC is false.

The Lumpl/Goliath puzzle

The final puzzle I want to look at here is known to contemporary philosophers as 'the Lumpl/Goliath puzzle'. Like the Ship of Theseus puzzle, the Lumpl/Goliath puzzle is a puzzle specifically about artefacts. However, the Lumpl/Goliath puzzle differs from all of the other puzzles that I examine here in that it does not have to do with the possibility of compound material objects remaining the same through time and change. The Lumpl/Goliath puzzle, rather, raises concerns about material constitution *at a time*. To see why, note that 'Lumpl' picks out a piece of clay and 'Goliath' a statue representing the giant of Biblical renown. By further stipulation, Lumpl and Goliath are always composed of the very same fundamental particles, and come into and go out of existence at the same time. However, Goliath and Lumpl appear to have different essential properties. Given these parameters, the PMC can be generated from the Lumpl/Goliath puzzle as follows (call this formulation of Lumpl/Goliath 'LG'):

(1)   Lumpl is a piece of clay [by stipulation].
(2)   Goliath is a statue [by stipulation].
(3)   Statues and pieces of clay have different persistence conditions [assumption].

(4) If objects x and y have different persistence conditions, then objects x and y have different essential properties [from the notions of a persistence condition and an essential property].

(5) If objects x and y have different essential properties, then objects x and y are non-identical [from the notion of an essential property and IMO6].

(6) Lumpl and Goliath are always composed of the same proper parts in possible world W [by stipulation].

(7) Lumpl and Goliath are not numerically identical in W [from (1)–(5) and IMO6].

(8) Lumpl and Goliath are numerically identical in W [from (6) and IMO5].

Why should one accept the premises of LG? Of course, premises (1) and (2) assume that there are such things as pieces of clay and statues (IMO1 and IMO2). Premise (3) is relatively uncontroversial among contemporary metaphysicians. An object's *persistence conditions* are the properties of an object that pick out the changes that it can and can't undergo and still remain in existence. But pieces of clay can survive changes that statues cannot, e.g. a statue cannot survive being melted but pieces of clay can – since the liquid clay can be composed of the same clay as the solid clay – whereas statues can survive changes that pieces of clay cannot, e.g. losing a part. Premises (4) and (5) are based on the notion that P is an essential property of x if and only if x has P in every possible world in which x exists. Therefore, as premise (5) has it, if objects x and y have different essential properties, then x and y are not identical. However, premise (5) also assumes that identity is transitive across possible worlds (IMO6). Premise (4) is based on the plausible assumption that an object's persistence conditions are numbered among that object's essential properties (although an object has more essential properties than its persistence conditions). Therefore, since Goliath and Lumpl have different persistence conditions, and therefore different essential properties and, assuming that identity is not relative to possible world (IMO6), Goliath and Lumpl *are not identical* in possible world W. But by stipulation Lumpl and Goliath are always composed of the same parts in world W. Assuming that spatially co-incident objects are impossible, Lumpl and Goliath therefore *are identical* in world W. The Lumpl/Goliath puzzle gives us another way to generate the PMC.

Aquinas's response to the Lumpl/Goliath puzzle is going to be reductionist in orientation: the premises of the argument wrongly assume that, like pieces of clay, statues are substances. Or alternatively, the puzzle wrongly assumes that pieces of clay and statues are on an ontological par with one another. Since Lumpl and Goliath are material objects in different senses (one is a material substance, the other is the same material substance considered under a certain phase sortal, 'statue'), it is no wonder that Lumpl and Goliath are thought to be both identical and non-identical. The sense in which Goliath and Lumpl are both identical and non-identical does not conflict with any laws of logic.

Although Lumpl has accidental forms at any time in which it exists, it need not have any of these in order to continue to exist. In contrast, since Goliath is an artefact, and therefore an accidental being, it does have an accidental form, e.g. *being in the shape of a giant*, as a necessary condition for its existence. As we saw in Chapter 5, the identity of an artefact over time requires sameness of artefact-configuring form (an accidental form). Goliath's identity over time should be thought of in the same way as any substance, *considered under a phase sortal* – such as the identity of a human child or seated-Socrates. Like the referents of 'human child' or 'seated-Socrates', Goliath is an accidental being, or being *per accidens* (through another), and Goliath's identity with Lumpl is only *per accidens*: the matter of Goliath (a substance) is numerically identical to Lumpl. In addition, the sense in which Goliath and Lumpl are non-identical – and therefore co-incident – is an innocuous one. Goliath's spatial co-incidence with Lumpl is no more troubling than the fact that Socrates, considered with a certain accident, e.g. *being seated*, is different from Socrates, considered without reference to that particular accident. Of course, the key to Aquinas's move here is his view that an artefact such as Goliath is not a material substance. The fact that Goliath and Lumpl are co-incident is thus non-problematic.

The advantage of a Thomistic approach to the PMC

As we have seen, contemporary philosophers argue that puzzles about compound material objects show that one must give up at least one common-sense intuition about material objects. This is what I have been calling 'the problem of material constitution' (PMC). What I want to suggest here is that Aquinas's metaphysic of material objects has the virtue of being able to *dissolve* the PMC. Given Aquinas's metaphysical commitments, one can solve the puzzles that apparently raise the PMC without giving up any of the following common-sense intuitions about compound material objects:

(IMO1) There are such things as *compound material objects*, that is there are material objects that are composed or constituted of other material objects.

(IMO2) There are many, many different kinds of compound material objects, including different kinds of atoms, molecules, aggregates of atoms or molecules, proteins, enzymes, plants, animals, tissues, organs, limbs, body sections, artefacts and artefact parts.

(IMO3) Compound material objects endure through time and change.

(IMO4) There are compound material objects that can survive certain losses, gains and replacements with respect to their parts.

(IMO5) Two material objects cannot exist in the same place at the same time.

(IMO6) Necessarily, identity is a transitive relation.

As far as Aquinas's commitment to these common-sense intuitions is concerned, his analysis of non-living and living compound substances shows that he is unconditionally committed to IMO1. Aquinas's view that a substance x's substantial form is sufficient to preserve x's identity through time and change – even changes with respect to x's integral parts – shows that he is also fully committed to the truth of IMO3 and IMO4. Even if what I called Aquinas's 'restrictive account' of artefact identity is correct, Aquinas is still not a mereological essentialist about artefacts. This is because an artefact x can survive those changes with respect to x's integral parts that the substances functioning as x's material quotient can survive (and, although non-living substances are very fragile, they can survive certain changes with respect to their parts for Aquinas). We have also seen that Aquinas is committed to IMO5; this despite the fact that, with respect to material substances, he accepts that composition out of integral parts is not the same as identity. Finally, in Chapter 5 I showed it is reasonable to assume that Aquinas is committed to IMO6. It is Aquinas's commitment to IMO2 that requires some clarification.

As I suggested in Chapter 1, espousal of IMO2 might be given the name *ontological pluralism*. It is Aquinas's commitment to ontological pluralism that makes the PMC such a live problem for his metaphysics. So perhaps it should be obvious that Aquinas does indeed accept IMO2. However, IMO2 is subject to more or less conservative versions and, although Aquinas would accept some of these, he would reject others. For example, take the following version of IMO2, one that Aquinas would reject:

(IMO2*) There are many, many different kinds of compound material *substances*, including different kinds of atoms, molecules, aggregates of atoms or molecules, proteins, enzymes, plants, animals, tissues, organs, limbs, body sections, artefacts and artefact parts.

Generally speaking, IMO2* captures Lynne Rudder Baker's own particular commitment to ontological pluralism.[17] In contrast to IMO2*, Aquinas's own commitment to ontological pluralism can be captured by the following:

(IMO2**) There are many, many different kinds of compound material *substances*, including many, many different kinds of non-living and living substances. There are also many, many different kinds of compound material objects that *do not* qualify as substances, such as the functional integral parts of plants and animals and the non-functional integral parts of non-living substances. Finally, there are objects that are compound material objects only in an extended sense, e.g. artefacts, artefact parts and aggregates of non-living substances.

Aquinas's commitment to IMO2** rather than IMO2* shows up in the way that he would handle the puzzles raising the PMC which explicitly deal with artefacts. Those puzzles wrongly assume that artefacts are on an ontological par with substances such as fundamental particles, pieces of bronze and living things. Aquinas thinks about artefacts in particular in much the same way that a reductionist thinks about compound material objects in general, that is as instances of a phase sortal rather than a substance sortal. Aquinas's approach to these puzzles is thus reductionistic.

Despite his reductionistic approach to artefacts, Aquinas should still be considered an ontological pluralist and an advocate of IMO2. To see why, compare Aquinas's views with those of Van Inwagen, who, strictly speaking, is a pluralist, although not an advocate of IMO2. We might say that Van Inwagen is a pluralist since he thinks that there are many, many kinds of material objects (namely, all varieties of living organisms). Nevertheless, Van Inwagen's general approach towards the PMC is eliminativist: he denies that there are artefacts, aggregates of particles, parts of material objects such as body sections, tails, heads and brains. In contrast with Van Inwagen, Aquinas thinks that there are functional parts of organisms such as tails, heads and brains and, arguably, Aquinas also believes in non-functional body sections such as Tib. However, Aquinas does not regard these types of material objects as material substances. Furthermore, Aquinas's way with aggregates and artefacts is not quite eliminativist – as with Van Inwagen – but rather reductionist: Aquinas thinks there *are* such things as aggregates and artefacts; they simply aren't substances but composites of substances and a certain sort of accidental form. Unlike Van Inwagen, Aquinas is surely committed to ontological pluralism, even if his own commitment to it is attenuated in comparison to Baker's.

As I have said, Aquinas still gets around the PMC where puzzles having to do with artefacts (the Ship of Theseus and Lumpl/Goliath) and aggregates (Debtor's Paradox) are concerned by denying that artefacts and aggregates are substances. So part of Aquinas's solution to the PMC is to deny that certain material objects, artefacts, exist in the way they are commonly thought to exist. But I think we also have a strong intuition that there is a fundamental ontological difference between artefacts and naturally occurring material objects such as organisms, namely that artefacts are not as real as naturally occurring material objects such as living organisms. And, of course, Aquinas's views are in accord with *that* intuition about artefacts.

But reductionism is not Aquinas's last word on the PMC; many of the puzzles I have looked at do not hinge on treating aggregates or artefacts as robust material objects (the Dead Debtor; the Tibbles-style puzzles).[18] Since, as we have seen, Aquinas does not deny *any* of the common-sense assumptions that generate the PMC for these puzzles, how can Aquinas solve them? Simply put, by admitting the distinction between *actual* and *potential* being.

In the puzzle of the Dead Debtor, Aquinas gets around co-incidence because the integral parts of Joe, which become integral parts of an aggregate after Joe's demise, are only potentially substances while they compose Joe. While these parts compose a substance such as Joe, they are not actual substances, but only the integral parts of a substance. Since Joe's parts have their being from Joe's substantial form, there is no sense in which Joe's parts compose an aggregate of actual substances that is co-incident with Joe. This, it seems to me, gives Aquinas's view a definite advantage over any view that fails to recognize the distinction between actual and potential being. Without that distinction in place, it is difficult to see how there cannot be an object co-incident with Joe, namely, the aggregate of Joe's parts. But if one supposes that there are no such material objects as aggregates, in that case it is still hard to see how Joe is not really just a derivative entity, given that his integral parts are all actual substances. If Joe's integral parts (the xs) are actual substances, it looks as though Joe turns out to be a mere derivative entity, something whose being merely supervenes on the xs.

Aquinas's evocation of the act/potency distinction also helps explain the metaphysical origin of the aggregate of substances that is identical with Joe's corpse. For Aquinas the aggregate of substances (call it 'the corpse's aggregate') identical with Joe's corpse *cannot* be identical to the set of Joe's integral parts, for the corpse's aggregate is newly generated upon Joe's death. But Aquinas's view here differs from the dichronicist who solves the puzzle of the Dead Debtor by saying that the change from Joe's integral parts to the corpse's aggregate lacks a substratum. For Aquinas, the substratum for this change – indeed, the substratum for any such substantial change – is prime matter, or being that is pure potency. Thus, Aquinas's metaphysics allows for a solution to the Dead Debtor puzzle that has the advantage of not sacrificing the common-sense view that all change requires a substratum.

As for the Growing Argument and the Tibbles-style puzzles, Aquinas can say much the same sort of thing. The material objects that cause the problem – the proper part of an object x that used to be identical in size to x, or the proper part of Tibbles that is all of Tibbles save her tail – are not actual substances, but merely potential substances, according to Aquinas. Thus, contrary to what these puzzles assume, Aquinas thinks that a proper part of a substance can never be identical to an actual substance, and Aquinas has a metaphysical explanation for this – substances are beings having the highest sort of unity, and so cannot be composed of substances themselves. Now in the Tibbles-style puzzles, it is simply assumed that a proper part can retain its identity when it is transformed into an improper part (or vice versa, in the case of the Growing Argument). But this assumption is not so much based on common sense as it is on crude observation (lots of things that look the same in fact are not – think of Joe and Joe's corpse right after Joe's untimely demise). Thus, Aquinas's rejection of this assumption is not incompatible with my thesis that Aquinas solves the PMC without denying any common-sense assumptions about material objects.

In closing, what I have wanted to suggest here is that Thomistic solutions to the PMC are novel and interesting since they solve the PMC without denying any of the common-sense assumptions about material objects that are thought to generate it. This is in direct contrast to those solutions typically advocated by contemporary philosophers, each of which entails that a common-sense intuition about material objects be rejected. Of course, Aquinas's solutions to Tibbles-style puzzles do entail buying into certain metaphysical assumptions that are not widely shared by our contemporaries. However, I hope to have shown in Chapters 3, 4, and 5 that Aquinas has good reasons for making such assumptions. Based as it is on a coherent and plausible metaphysic of material objects, the Thomistic solutions to the PMC have the advantage of being more in accord with common-sense intuitions about material objects than the most important contemporary solutions. I now turn in the final chapter to defending and developing this claim in greater detail by responding to some objections to Aquinas's metaphysic of material objects.

Notes

1. In fact, as we saw in Chapter 5, Aquinas thinks that the Ship of Theseus loses its numerical identity upon losing only one of its planks since an artefact's form of the whole preserves that artefact's identity through time and change and artefact forms are individuated by the matter they configure (which in the case of the Ship of Theseus is an aggregate of pieces of wood, and aggregates are in turn individuated by their *members*). However, I do not need to rely on Aquinas's restrictive account of artefact identity here. Even Baker, who thinks that artefacts can survive the replacement of some of their integral parts, agrees that an artefact cannot remain numerically the same if it undergoes total part replacement (Baker 2000, pp. 36–7).

2. See IA in Chapter 5.

3. I take the term from Van Inwagen 1990a, pp. 74–80.

4. For an interesting defence of the falsity of IE, see Van Inwagen, P. (1998) *The Possibility of Resurrection and Other Essays in Christian Apologetics*, Boulder, CO: Westview Press, pp. 43–52.

5. See, e.g., Rea 1997a, xvi–xvii.
6. Rea 1997a, xxviii.
7. Aquinas might also reject premises (1) and (3) on the basis of being ill-formed. As we saw in Chapter 5, Aquinas thinks that identity statements have substances as their *relata* and, whatever aggregates are for Aquinas, they are not substances. So 'Socrates at *t* is identical to aggregate$_{at\ t}$' is an ill-formed proposition according to Aquinas.
8. The dominant-kinds view espoused by Burke (1994a) and Rea (2000) therefore *resembles* the Thomistic solution to DP (see Chapter 1 for some discussion of the dominant-kinds view). However, Aquinas's metaphysic of material objects provides a solution to a problem that plagues the dominant-kinds view, namely, the 'which one' problem. Why think there are human organisms existing wherever Chris and Merry are and not just aggregate-objects? For Aquinas, the presence of a unitary substantial form where Chris and Merry are – and with it whole classes of properties not had by a mere aggregate-object – explains why it is sensible to say that there are human organisms and not aggregate-objects existing wherever Chris and Merry are. See Chapter 7 for further discussion of a substantial form's ability to confer whole classes of properties on an object not had by a mere aggregate-object.
9. Of course, one might also deny the identity claims in (1) and (3) by denying IMO3, as the temporal-parts theorist does. I ignore this possibility for the purposes of the discussion here.
10. Doepke, F. (1997) 'Spatially coinciding objects' in M. Rea (ed.), *Material Constitution: A Reader*, Lanham: Rowman and Littlefield, p. 11. The reader should note that Doepke himself does not endorse the dichronic view.
11. See, e.g., Van Inwagen 2001, p. 108.
12. For some discussion of the historical origins of the Growing Argument, see: Sedley, D. (1982) 'The Stoic criterion of identity', *Phronesis* 27: 255–75.
13. See my discussions of Van Inwagen's views in Chapter 2.
14. For some remarks on the history of this argument, see Rea 1997a, xviii.
15. See Chapter 4 for the distinction between functional and non-functional integral parts.
16. Van Inwagen 1990a, p. 172.
17. Although Baker does not talk in terms of a substance-ontology, I believe she is generally committed to such a view. At the very least it does not seriously misrepresent Baker's views to say that she thinks that artefacts have the ontological status of substances.
18. Furthermore, one might want to suggest that Aquinas's position on artefacts could be amended so as to allow artefacts the ontological status of substances. In that case, Aquinas's solutions to puzzles involving artefacts will parallel his solutions to the Tibbles-style puzzles. Just as Tib goes out of existence when Tibbles loses her tail, so a non-artefact, non-living substance will lose its identity when it becomes the matter out of which an artefact is made. Again, compare this solution to the dominant-kinds view of Burke 1994 and Rea 2000.

7

Answering Objections to the Thomistic Approach to the PMC

In the last chapter I argued that Aquinas's metaphysic of material objects allows him a solution to the PMC that has an important advantage over most contemporary theories: puzzles about compound material objects can be solved without denying *any* of the common-sense intuitions about compound material objects that are supposed to generate the PMC. In this chapter I want to solidify my argument by answering potential objections to the Thomistic solutions coming from the contemporary philosophers whose views and solutions to the PMC I sketched in Chapter 2: those of Lynne Rudder Baker, Peter van Inwagen and Dean Zimmerman.

Aquinas and Baker: the advantages of a mitigated ontological pluralism

Aquinas and Lynne Rudder Baker are both ontological pluralists about material objects. They both think that there are many, many different kinds of compound material objects; some are non-living while others are alive, and some belong to natural kinds, while others are artificial. In this regard, they both develop a metaphysic of material objects that is anti-reductionist in character.

However, Baker and Aquinas defend different versions of ontological pluralism. Baker gives every indication that instances of natural kinds and artificial kinds should be treated on an ontological par. In contrast, Aquinas sees a major ontological difference between the instances of natural kinds and the instances of artefact kinds. Aquinas admits that artefacts are individual objects in some sense. Nonetheless, he thinks about artefacts reductively: an individual artefact is not an instance of a substance-sortal but rather an instance of a phase-sortal. Artefacts are composites of a substance (or group of substances) and an accidental form and are not substances themselves, according to Aquinas.

I imagine that Baker might object to Aquinas's position on the ontological status of artefacts by saying that it is rather counter-intuitive.

We certainly have an intuition – which shows up in our social and legal practices – that artefacts are material objects on an ontological par with the instances of natural-kinds. Aquinas's metaphysic of material objects is in conflict with this intuition. Therefore, Aquinas offers a counter-intuitive approach to material objects.

Call this the 'Baker Objection' to Aquinas's metaphysic of material objects. The Baker Objection raises a concern for the viability of Aquinas's metaphysics, but it is also a challenge to my argument that the Thomistic solutions to the PMC hold an advantage over contemporary solutions such as Baker's.

In responding to the Baker Objection, I first want to point out that it is not at all clear that Aquinas's position on the ontological status of artefacts is an essential part of his metaphysic of material objects. The Thomist could reject Aquinas's views on artefacts without doing any damage to Aquinas's concept that a substance is a unified material object of the highest sort and his doctrine that substances are composed of both metaphysical and integral parts, none of which count as substances themselves.

However, it seems to me that the Baker Objection is wrong to assert that it is counter-intuitive to deny that artefact objects are just as real as the instances of natural kinds. In fact, don't we have the opposite intuition – whatever our practices might suggest? On an intuitive level, do we really think that bikes, computers and other machines enjoy the same sort of ontological status as carbon atoms and cats, let alone creatures such as ourselves? Rather, I think we have the intuition that artefact objects are aggregates that we treat as unified objects.

In addition, Aquinas's account of artefacts as mere accidental beings can be defended by looking at the counter-intuitive implications of a view on artefacts such as Baker's, one that treats artefacts as substances. According to Baker's constitution view, artefacts enjoy the same ontological status as instances of natural kinds.[1] But, according to Baker, an artefact does not belong to a substance sortal for the same reasons that an instance of a natural kind does. An instance of a natural kind belongs to a substance sortal in virtue of its 'intrinsic properties', or in virtue of the relations that obtain between its integral parts. In contrast, an artefact belongs to a substance sortal – or, in Baker's terms, 'a primary kind' – in virtue of that artefact's enjoying certain relations to a community of rational agents who think about and treat that object in certain ways, e.g. with reverence, respect or perhaps simply *as an object*. Whereas Aquinas thinks that a statue is a composite of a non-living substance and a certain sort of intrinsic property – an accidental form such as a certain sort of shape – Baker takes an artefact to be a 'substance' in its own right, and this in virtue of a relation to some object extrinsic to the piece of matter that constitutes the artefact.

By granting that artefacts have substantial existence because they have essential properties in virtue of relations to objects extrinsic to them, one has to admit that there are nearly an infinite number of different primary kinds having instances. The ontology of Baker's constitution view is therefore not only bloated, but is in a constant state of *ballooning*: there are in fact new primary kinds and new instances of many (rather odd) sorts of primary kinds coming into existence all the time.

Baker admits that she does not have a principled way of explaining why some properties are primary-kind properties while other properties are ones that objects belonging to primary kinds have only accidentally. As she notes, this should not be at all surprising, since such 'a theory of primary kinds would be tantamount to a theory of everything'.[2] Nevertheless, as a second-best alternative, Baker offers the following principle – I'll call it 'P':

(P) if *x* constitutes *y*, then *y* has whole classes of causal properties that *x* would not have had if *x* had not constituted anything.[3]

Baker thinks that this principle helps us to see that, for example, *being a husband* is not a primary-kind property. According to Baker, this is because husbands do not have whole classes of properties that human beings who are not husbands do not have. However, this does not seem right. There *are* whole classes of causal properties that typically belong to husbands that do not belong to bachelors. Consider just those causal powers that a husband has because of our legal conventions. For example, a husband causes his offspring to be legitimate, and his wife will be the heir of a sizable portion of his assets when he dies.[4] Given the fact that a human being does have whole classes of properties in virtue of being a husband, if we are to treat 'statue' as a primary kind, why not 'husband'?

Of course, Baker is right to reject that *being a husband* is a primary-kind property. A human being does not come to constitute another individual object simply because he has causal powers in virtue of certain legal conventions. *Being a husband* is not a primary-kind property because the causal powers of a husband are rooted in something *extrinsic* to the human being who happens to be a husband.

Given the parallel between the properties *being a husband* and *being a statue*, just as we should reject that *being a husband* is a primary-kind property, we should also reject that *being a statue* is a primary-kind property. In order for an object to exist in the way that pieces of marble, parrots and people do, it must, like these objects, not depend for its existence on the presence of certain sorts of intentional states in a human being or beings. The problem with *statue* being a primary kind thus lies not in principle (P), but in Baker's insistence that an object *y* can have *primary-kind* properties in virtue of objects *extrinsic* to *y*, in this case those human beings who, following conventions, confer upon statues certain sorts of value.

If one admits with Baker that statues exist in just as robust a sense as snakes, then the list of counter-intuitive primary kinds that will have actual instances will be potentially endless. Insofar as I take some object to serve a purpose, with the possible additional constraint that I communicate this intention to others, it remains difficult to see why such an intention is not sufficient to bring about the existence of a new individual material object, since such an object 'has whole classes of causal properties that [a constituting object] would not have had if [that constituting object] had not constituted anything'.[5] For example, suppose I inform a friend, before going to a party with him, that, when I wave a paper napkin in the air, this should be a signal to him that I want to go home (say we are attending a socially obligatory event and I don't want to draw widespread attention to the fact that I am leaving the party early for no other reason than that I don't want to miss Monday Night Football). Given Baker's criterion for something belonging to a primary kind, doesn't the napkin – when I start thinking about it in a way that has pragmatic consequences – come to constitute a numerically different object, namely, an object having the primary-kind property *being a signal*? Or suppose that Joe is at a picnic, and wants to play a game of softball with his friends. In order to play softball, one needs to have objects that function as bases. To this end, Joe grabs four spare paper plates. At the time Joe gets the thought to use the paper plates to function as bases, do the paper plates come to constitute new objects? But surely one cannot just *think* individual material objects that have the same ontological status as living organisms into existence – at least it is not within the providence of human beings to do such things.

The problem I have been discussing here lies in treating artefacts as though they belong to substance instead of phase sortals. Since Aquinas's version of ontological pluralism does not entail such a commitment, it is to be preferred to Baker's. This is because Baker's view is committed to the counter-intuitive implication that there are nearly an infinite number of individual objects in the world belonging to nearly an infinite number of primary kinds, not to mention a commitment to the existence of instances of rather odd primary kinds such as 'husband', and 'signal-napkin'. Since Aquinas does not think about artefacts as substances, his version of ontological pluralism is not afflicted by such a ballooning ontology. Therefore, rather than being in conflict with common sense as the Baker Objection contends, Aquinas's reductionistic view of artefact existence is compatible with our intuition that artefacts are not objects in the same sense as are the instances of natural kinds. In fact, Aquinas's views on artefacts are part of a version of ontological pluralism that is more plausible than Baker's ever-ballooning constitution view.

Aquinas and Van Inwagen: defending the reality of non-living material substances

I have argued that Thomistic solutions to the PMC enjoy an advantage over those solutions offered by most contemporary philosophers: in contrast to these contemporary solutions – of which Peter van Inwagen's is a notable example – the Thomistic solutions can solve the PMC without giving up any of those common-sense assumptions about material objects that are thought to generate it. I now raise and attempt to answer an objection to the viability of Aquinas's metaphysic of material objects that, if successful, would severely weaken this argument for the superiority of Aquinas's views. The objection is one that I imagine Van Inwagen might raise:

> The Thomistic solution to the PMC has it that we can accept all of the common-sense intuitions about compound material objects that are thought to generate it. But one of these intuitions (IMO2) entails that there are non-living compound material objects. However, as I have argued for at length in 1990a, there is no principled way of showing that the xs compose y, where y is a *non-living* compound material object. Therefore, anyone who takes the Special Composition Question (SCQ) seriously – as I do – cannot accept the Thomistic solution to the PMC, given that it unjustifiably assumes the existence of non-living compound material objects.

As we have seen, Van Inwagen's proposed answer to the SCQ has it that the xs compose y if and only if the activity of the xs constitutes a life. Now this answer to the SCQ is too restrictive as far as Aquinas's allegiance to ontological pluralism is concerned. Although living compound material objects are certainly the most obvious and unproblematic examples of compound material substances, there are also many kinds of non-living compound material substances in the world by Aquinas's lights. Furthermore, there are other sorts of compound material objects that Van Inwagen's proposed answer to the SCQ does not save – artefacts, aggregates and arbitrary undetached parts – all objects that Aquinas takes to have at least *some* measure of ontological significance. However, I shall merely focus here on providing an answer to the SCQ that allows for the existence of non-living *substances* and therefore answers the Van Inwagen-style objection that I raised above.

Recall that, for Aquinas, the species of an individual substance is given by its substantial form. So one might propose the following Thomistic answer to the SCQ:

> (SF) The xs compose a substance y if, and only if, (a) there is a z that has a *substantial form* such that z belongs to a species different from any of the xs

and (b) z and the xs are spatially co-incident, and (c) the xs have their being and species in virtue of z's substantial form, and (d) z is numerically identical to y.[6]

SF gives us one important piece of a distinctively Thomistic answer to the SCQ. But we need more. This is because one might reasonably ask under what sorts of conditions there is a z that has a substantial form such that z belongs to a species different from any of the xs (where z and the xs are spatially co-incident and the xs have their being and species in virtue of z's substantial form). I think Baker's principle (P) and the contemporary notion of an emergent property can help provide a good answer to this question.

First, I need the concept of an emergent property. Consider the following definition of an emergent property of a system, offered by John Searle:

> An emergent property of a system is one that is causally explained by the behavior of the elements of the system; but it is not a property of any individual elements and it cannot be explained simply as a summation of the properties of those elements.[7]

Elsewhere Searle suggests that there is 'a much more adventurous conception' of emergence than the one offered above.[8] This 'more adventurous' conception of emergence rejects the first part of the definition above – that the emergent property 'is causally explained by the behavior of the elements of the system' – but accepts the second part – that an emergent property 'is not a property of any individual elements [in the system] and it cannot be explained simply as a summation of the properties of those elements [in the system]'.

Now consider the following passage from ST, where Aquinas makes the point that compound material substances have powers that are not caused by the powers of the kinds of things that compose those compound material substances. Such features thus seem to be emergent in Searle's 'more adventurous' sense:

> It must be considered that the more noble a form is, the more it rises above corporeal matter, and the less it is merged in matter, and the more it exceeds matter by its operation or power. Hence, we see that the form of a mixed body has a certain operation that is not caused by [its] elemental qualities. And the more one proceeds in the nobility of forms, the more it is found that the power of the form exceeds the elementary matter, just as [the power of] the vegetative soul is more than the form of a metal, and the sensible more than the vegetative soul. But the human soul is the highest in the nobility of forms. Hence, it exceeds corporeal matter in its power such that it has a certain operation and power in which in no way does it communicate with corporeal matter. And this power is called the intellect.[9]

A composed substance – because of its substantial form – has properties that none of the integral parts of that substance has, whether those integral parts are considered *singillatim* or as a sum. Thus, the properties unique to the instances of a particular substance-kind might be spoken of as 'emergent' for Aquinas. And because, as Aquinas says in the passage above, 'the form of a mixed body is not *caused* by [its] elemental qualities', it appears as if Aquinas thinks that the properties of a compound material substance are emergent in Searle's more adventurous sense of that term.[10]

With this notion of emergent property in place, recall Baker's principle for explaining when it is the case that an object *x* constitutes an object *y*:

(P) if *x* constitutes *y*, then *y* has whole classes of causal properties that *x* would not have had if *x* had not constituted anything.[11]

Taking principle P and the notion of an emergent property as points of departure, I want to suggest that the Thomist can explain what it means for *y* to have a substantial form such that *y* belongs to a species different from *y*'s integral parts (the *x*s) in the following way (call it 'CP' for 'whole classes of *causal properties*'):

(CP) *z* has a *substantial form* such that *z* belongs to a species different from the species of any of the *x*s, where *z* and the *x*s are spatially co-incident and the *x*s have their being and species in virtue of *z*'s substantial form if, and only if, *z* has whole classes of intrinsic causal properties that the *x*s, whether taken *singillatim* or as a sum, do not have.

Given SF and CP, let me thus propose the following Thomistic answer (TA) to the SCQ:

(TA) The *x*s compose a substance *y* if, and only if, (a) there is a *z* that has whole classes of intrinsic causal properties that the *x*s, whether taken *singillatim* or as a sum, do not have, (b) *z* and the *x*s are spatially co-incident, (c) the *x*s have their being and species in virtue of *z*'s substantial form, and (d) *z* is numerically identical to *y*.

One advantage that TA has over Van Inwagen's proposed answer to the SCQ is that TA is compatible with the existence of non-living composed objects. To take a contemporary example, consider the case of a water molecule, an object composed of two hydrogen atoms and an oxygen atom. The water molecule has whole classes of intrinsic causal properties that hydrogen atoms and oxygen atoms do not have, taken individually (and even when the properties of those atoms are considered as a sum). For example, the water molecule has the property *being able to join with other water molecules to form a potable liquid that all living organisms require in generous amounts in order to continue functioning.* Oxygen atoms and hydrogen atoms do not have this property; nor will the mere summation of the properties of hydrogen and oxygen atoms produce such a property. That water molecules have this property means that each water molecule has whole classes of intrinsic properties that its integral parts do not have, and so, according to TA, a water molecule is a composed object – a compound material substance with its own substantial form – and not merely an aggregate of substances sharing an accidental form. To take another example, every human being has whole classes of intrinsic causal properties that their integral parts do not have, e.g. *being able to understand* and *being able to will some action freely.* Human beings have causal powers that are not shared by any of the integral parts of a human being, taken *singillatim*, nor are these distinctively human powers identical to a summation of the powers and properties of a human being's integral parts. A human being has whole classes of properties that a human being's integral parts do not have. Therefore, by TA, every human being is a composed material substance.[12] Similar sorts of arguments could be given for a wide variety of non-living and living material objects.

Note that TA has it that the *x*s compose *y* only if *y* has whole classes of causal *intrinsic* properties that the *x*s, whether taken *singillatim* or as a sum, do not have. Thus, TA is compatible with the plausible view that the *x*s cannot compose *y* merely as a result of the *x*s coming to have relations to objects *extrinsic* to the *x*s. For example, if an artefact or artwork is an object that exists only because pieces of matter, e.g. an aggregate of bricks, instigate certain thoughts in the minds of a community of rational agents, thoughts such as 'these bricks, arranged in a certain way, make a useful and beautiful structure', then an artefact or artwork does not satisfy TA. If being thought about in a certain way is a property of the bricks, it is only an extrinsic or Cambridge property of the bricks much like Socrates's property *being classified by historians of philosophy as the Greek philosopher that divides the Greek philosophers primarily interested in natural philosophy from those Greek philosophers interested primarily in ethical matters.* This property is one Socrates has not intrinsically, but extrinsically. Contra Baker, Aquinas thinks that a substance's essential properties cannot be properties of this sort.

TA thus offers a principled way of deciding when it is the case that the *x*s compose a material substance. Of course, TA won't satisfy everyone. Some philosophers will surely regard those properties I claim could only belong to a composed object as mere 'network properties', that is, properties had by the *x*s that arise when the *x*s enter into certain novel causal and/or spatial relations.[13] According to such philosophers, any properties of so-called compound objects may also be construed as being merely network properties of the *x*s (and not some individual material object *y*). Now Van Inwagen defends his own solution to the SCQ by relying on the assumption that thinking could not be a cooperative activity. Thinking is thus not an activity analogous to light shining, for example. A plurality of objects could be the cause of light shining – as in the case of light emanating from the sun (the sun is not an individual material object for Van Inwagen). But a plurality of objects could not be the cause of the activity of thinking; thinking requires a unified subject.[14] Of course, a reductionist or eliminativist about the mind could simply say that thinking is a network property of the xs, and so is, after all, a cooperative activity. If TA therefore has a weakness, then it is the same sort of weakness shared by Van Inwagen's solution to the SCQ: at the end of the day, such a solution won't satisfy the sceptic.

Nevertheless, in contrast to Van Inwagen's proposed answer to the SCQ – the xs compose y if the activities of the xs constitute a life – TA is compatible with IMO2: it allows that there are not only living compound material objects, but non-living ones as well. Furthermore, TA justifies belief in living and non-living compound material objects without also admitting that artefacts enjoy the same ontological status as material objects such as atoms, molecules, enzymes and living organisms. Since TA answers in a principled way when it is that the xs actually compose something, and in a way that is compatible with (an intuitively plausible version of) ontological pluralism, I conclude that TA successfully answers the Van Inwagen-style objection to my Thomistic solution to the PMC.

The Zimmerman Argument: a Thomistic response

According to the Zimmerman Argument (ZA), many of the objects that we commonly take to be composed *material* objects are in fact abstract objects, events or even mere virtual objects. What makes ZA interesting for the Thomist is that it takes for granted so many common-sense intuitions about material objects: individual material objects endure through time (IMO3), two individual material objects cannot be wholly spatially co-incident (IMO5) and, if individual material objects x and y are identical, then necessarily, x and y are identical (IMO6).

Recall that ZA has roughly the following form. IMO1, IMO3, IMO5 and IMO6 are all true. If IMO2 is true, then IMO4 is true (for then there would be at least some kinds of compound material object able to survive part replacement). But if IMO4 is true, then IMO5 is false (this is because constituted objects – objects that can survive replacement of their parts – would be spatially co-incident with non-atomistic homeomerous masses of K – objects that cannot survive the replacement of their parts). Therefore, it is not the case that IMO2 is true. ZA turns the PMC into an argument for the falsity of ontological pluralism.

Like many of the arguments we have been looking at that have to do with the PMC, ZA has bite only for the philosopher who accepts certain intuitions or assumptions about material objects. So, for example, ZA assumes that material objects endure through time (IMO3). The philosopher who favours the doctrine of temporal parts does not have to worry about ZA – since temporal-parts theorists reject IMO3. However, there are other assumptions and intuitions at play in ZA besides IMO1, IMO3, etc. These include the following: (1) There are constituting masses, that is, entities constituting objects such as tigers and tables that have persistence conditions such that they are essentially related to their parts. (2) These constituting masses are not identical to the objects they constitute. (3) Atomism is false, or at least possibly false. Now, given what we have seen in Chapters 3, 4 and 5, I think Thomas Aquinas accepts (at least slight variations of) all of these assumptions. So ZA constitutes a genuine challenge to a Thomistic solution to the PMC in that it tries to show that (1)–(3) are incompatible with IMO2. Thus, in order to vindicate my Thomistic solution to the PMC, I need to show that Aquinas's metaphysics allows him a way of dissolving ZA. Having done so, my argument that a Thomistic solution to the PMC is preferable to contemporary solutions (including Zimmerman's) will receive further support. But does Aquinas's metaphysic of material objects allow him a way of dissolving ZA?

Aquinas's approach to the nature of material objects constitutes a kind of multiple-category approach to material constitution that ZA overlooks, one that *is* compatible with ontological pluralism and the truth of either atomism or non-atomism. As Zimmerman writes, 'I believe that the only plausible multiple-category theories … are … those identifying constituted objects with processes, or with logical constructions out of masses. These are, at the very least, the *only* other multiple-category metaphysics that anyone has proposed.'[15] I propose that a Thomistic approach to compound material objects is a plausible multiple-category approach that does not entail that constituted objects are mere processes or logical constructions. Rather, constituted objects such as trees and tigers are what we intuitively think they are: garden-variety compound material objects.

Consider a theory of masses that has the following focus: whether certain masses are substances or non-substances. Call such a theory of masses, 'a substance theory of masses'. According to the substance theory of masses, some mass expressions refer to individual material substances, whereas other mass expressions do not, either because the entities to which such expressions refer are individual material objects of a different ontological category than 'substance', or because those mass expressions refer – as in the plurality theory of masses – to a plurality of individual material substances. Given what we have seen in Chapters 3, 4, and 5, Aquinas's metaphysic of material objects can be construed as a 'substance theory of masses'.

A substance theory of masses would appear to qualify as a kind of – to use Zimmerman's expression – 'hybrid' theory of masses, that is a theory of masses where mass expressions sometimes refer to individual material objects, and sometimes not.[16] A hybrid theory of masses contrasts with both the sum theory of masses (which takes every mass expression to refer to an individual material object) and the plurality approach to masses (which takes every mass expression to refer to a mere plurality of individual material objects). Although Zimmerman mentions the possibility of a hybrid approach to understanding the referents of mass expressions briefly in one place, he appears not to take it very seriously (and without any argument).[17]

That Aquinas takes some mass expressions to refer to a plurality of material substances is clear from the fact that he would take a locution such as 'God made all of the water existing in the world' to be a perfectly meaningful expression. Now 'all of the water in the world' is a mass expression that certainly cannot be converted into some ordinary count noun and so refers to a mass of water. But Aquinas does not think that the referent of this expression is a *parti-cular* substance. A scattered object – like the referent of 'all the water existing in the world' – is not a particular substance for Aquinas. (How could, for example, the sum of the water in Lake Michigan, Lake Superior, the Pacific Ocean, etc. be a *particular* substance?) But, on the other hand, I see no reason why Aquinas could not say that 'all of the water existing in the world' refers to a plurality of material substances, namely all those particular instances of the species 'water' that count as particular substances at any given time.

In addition to some mass expressions referring to pluralities of material substances, the substance theory of masses also has it that some mass expressions refer to particular material substances. Examples might include the referents of 'the water in the puddle' and 'the bronze that composes the statue'. Since Aquinas takes some mass expressions to refer to individual material substances and other mass expressions to refer to mere pluralities of material substances, Aquinas's substance theory of masses can also be considered a hybrid theory of masses.

But Aquinas's hybrid theory of masses is rooted in his own substance ontology. One aspect of Aquinas's theory of substance that is important for translating it into a theory of masses is his view that substances cannot be composed of substances. If a substance were composed of substances, this would violate the principle that a substance is something actually one and not many. Therefore, on the substance theory of masses, masses that are substances are not composed of masses that are substances.

Recall that on the sum theory of masses, a mass x of K is a mereological sum of other masses of K. Is x a mereological sum of its proper parts, each proper part of x being identical to some K, according to the substance theory of masses? Of course, the answer is 'no'. If x itself were a mereological sum of K, then *x*'s proper parts would be substances that are proper parts of a substance, which Aquinas thinks is impossible.

However, this is not to say that a mass that is a substance cannot ever be a proper part of another individual material object. As we have seen, Aquinas thinks that aggregates have ontological significance. Not only heaps of stones, but artefacts too count as a kind of aggregate for Aquinas.[18] Thus, in Aquinas's view it is the case that a mass term might refer to an individual material object that is part of another individual material object, but only when 'individual material object' does not have the same meaning in both cases, e.g. an axe – an individual material object qua artefact – is composed of some metal and some wood – individual material objects qua material substances.

Finally, according to Aquinas's substance theory of masses, there are some mass expressions that do not refer either to a plurality of material substances, or to a particular material substance or to an aggregate of material substances. To see why, note that for Aquinas a mass *x* of *K* that is a substance can have proper parts that are *K*, where all of *x*'s proper parts are not substances but *integral parts of substances* as long as they remain proper parts of *x*. For example, consider that the referent of 'the water in my glass' is a substance for Aquinas whose integral parts are also masses of water. As Aquinas sometimes says, a substance *x*'s integral parts are potentially substances but not actual substances as long as they remain integral parts of *x*. As we have seen, Aquinas's views here revolve around his understanding of substance as something that is one in the highest sense. Therefore, Aquinas thinks that the integral parts of a substance owe their existence and identity to the substance to which they belong as parts.

These views of Aquinas's allow him to countenance the *logical* possibility that an instance of a homeomerous substance-kind – an element-kind for Aquinas – could be divided ad infinitum. Take the referent of 'the water in the tub' (assume for the sake of argument that water is a homeomerous – an elemental – kind). This mass of water is a substance for Aquinas. The mass of water could be divided in half, and the result would be two numerically different substances, both belonging to the stuff-kind 'water' and both being numerically different from the original mass of water. Presumably, this process could be continued without limit (if only in another possible world – since Aquinas thinks that the species *water* does in fact have *minima*). The important point here is that Aquinas's substance theory of masses is logically compatible with non-atomism and atomism.

Having sketched Aquinas's approach to masses – the substance theory of masses – let us return to the claims of ZA. Recall that it is specifically a theory of masses committed to treating masses as pluralities that falls prey to ZA's conclusion that constituted objects cannot be individual material objects. This is because, as ZA shows, if masses are identical to pluralities of material objects, then atomism is true. But on the supposition that non-atomism is true (which, as Zimmerman argues, is a real possibility), a plurality approach to masses would be false. But, if masses are not pluralities, they must be individual material objects, as the sum theory supposes. On the sum theory of masses, the homeomerous masses w, x, and y that constitute, say, an organism, are identical to a mass z that is the mereological sum of w, x and y. This is true even if w, x and y belong to different stuff-kinds, e.g. quark-stuff, electron-stuff, etc. But since z too is a mass, then it is an individual material object. Since spatially co-incident objects are not to be countenanced, the object that z constitutes, say a tiger, cannot itself be an individual material object. A tiger is either a virtual object – in that case tigers don't really exist after all – or else it is an event or process that is 'happening' to a mass of matter, akin to a wave moving through the water.

However, in contrast to the sum theory of masses, a substance theory of masses allows for the possibility that objects constituted of masses are individual material objects – even material substances – and does so without admitting the possibility of spatially coincident objects. This is because, according to the substance theory of masses, one ought to reject the assumption on the part of the advocate of the sum theory of masses that the summation of any two masses x and y is itself a mass z, where x, y and z all belong to the same ontological category, e.g. 'material substance'. According to the substance theory of masses, masses x and y are individual material substances only if they do not compose a mass z that is a *substance*. Therefore, masses x and y compose a substance, e.g. a human being, only if x and y are not actual substances, but merely substances *in potentia* or, alternatively, the integral parts of a substance. A substance and its parts belong to *different ontological categories*.

Zimmerman himself admits that there is no problem of spatial coincidence for multiple-category theories of material constitution. Since a theory of material objects that has it that a constituted object is composed of a plurality of material objects qualifies as a multiple-category approach to material constitution,[19] a fortiori Aquinas's view that a material substance is composed of a plurality of substances *in potentia* also counts as a multiple-category approach.

But this means that Aquinas's approach to masses allows him a way of dissolving ZA. This is because Aquinas's substance theory of masses is not, like the pluralities approach to masses, committed to atomism. Furthermore, Aquinas's hybrid substance theory of masses is not, like the sum theory of masses, committed to the view that masses are always individual material objects, *simpliciter*: some masses are substances, some are pluralities, some are aggregates and some are substances *in potentia*, or the integral parts of substances. Thus, assuming the viability of Aquinas's metaphysic of material objects, the substance theory of masses that is based upon such a metaphysic constitutes an alternative theory of masses, one that allows the ontological pluralist to block the unpalatable conclusion of ZA.[20]

Most contemporary solutions to the PMC entail the rejection of at least one common-sense intuition about compound material objects. I have argued that Aquinas's metaphysic of material objects enables him to solve the PMC without abandoning any of these intuitions. Since, all things being equal, a solution to the PMC that saves all of our common-sense intuitions about material objects is preferable to one that does not, I conclude that the Thomistic solutions to the PMC are preferable to any of the contemporary solutions that I have examined. Furthermore, based as they are on a coherent and plausible metaphysic of material objects, the Thomistic solutions are no ad hoc solutions to the PMC. Aquinas's metaphysic of material objects thus constitutes a powerful alternative to contemporary approaches to compound material objects, one that offers an able explanation and defence of our common-sense intuitions about compound material objects.

Notes

1. Indeed, some of the things Baker says imply that some artefacts have a greater ontological status than some instances of natural kinds. See Baker 2000, p. 25.
2. Baker 2000, p. 40.
3. Baker 2000, p. 41.
4. I am grateful to Eleonore Stump for the idea of focusing here on a husband's legal status, as well as for ideas about the kinds of causal powers that a husband has in virtue of that status.
5. Baker 2000, p. 41.
6. The reader should note that y and the xs are spatially co-incident only in the very innocuous sense of one–many co-incidence (see Chapter 6 for further discussion of one–many co-incidence), not to mention the fact that the xs are only potential and not actual substances at any time in which they compose y. Also, I do not intend the answer to the SCQ I develop in this section to be able to cover all cases of material composition in Aquinas. Specifically, I take it to be an answer to the SCQ where the composed object is a mixed body composed of elements, or a living organism composed of mixed bodies and elements. As we have seen, instances of an element-kind K can also be composed of integral parts belonging to K according to Aquinas and my Thomistic solution to the SCQ does not cover such cases.
7. Searle, J. (1997) *The Mystery of Consciousness*, New York: The New York Review of Books, p. 18. This is, of course, only one way of understanding an emergent property among many in the contemporary literature. For some good discussions of the different ways of understanding the notion of an emergent property, see, e.g., O'Connor 1994 and Hasker 1999, pp. 170–8.
8. Searle 1992, p. 112.

9. ST Ia. q. 76, a.1, c.: 'Sed considerandum est quod, quanto forma est nobilior, tanto magis dominatur materiae corporali, et minus ei immergitur, et magis sua operatione vel virtute excedit eam. Unde videmus quod forma mixti corporis habet aliquam operationem quae non causatur ex qualitatibus elementaribus. Et quanto magis proceditur in nobilitate formarum, tanto magis invenitur virtus formae materiam elementarem excedere: sicut anima vegetabilis plus quam forma metalii, et anima sensibilis plus quam anima vegetabilis. Anima autem humana est ultima in nobilitate formarum. Unde intantum sua virtute excedit materiam corporalem, quod habet aliquam operationem et virtutem in qua nullo modo communicat materia corporalis. Et haec virtus dicitur intellectus.' See also QDSC q. un., a. 2, c. and In Met. VII, lec. 17, nn. 1673–4.

10. I owe the idea of using Searle's ideas on emergence to explain Aquinas's non-reductive account of substances to Stump 2003, pp. 194–7.

11. Baker 2000, p. 41.

12. As we saw in Chapter 3, even if each human is composed of a part that is immaterial such that it can exist apart from matter, human beings are nonetheless rightly classified as material beings for Aquinas.

13. See, e.g., Churchland, P. (1986) *Neurophilosophy: Toward a Unified Science of the Mind/Brain*, Cambridge, MA: The MIT Press, pp. 324–5.

14. Van Inwagen 1990a, pp. 115–23. Van Inwagen's defence of his answer to the SCQ continues as follows: human beings are thinking beings and material beings. But thinking requires a unified subject. Therefore, human beings are compound material objects. But human beings are not essentially thinking things, and it is plausible to think that *being alive* is rather the feature of human beings that requires that the xs compose something in their case. Since it would be arbitrary to say that human beings exist but no other organisms do, Van Inwagen concludes that all living organisms are individual compound material objects.

15. Zimmerman 1995, p. 107.

16. Zimmerman 1995, p. 60.

17. Zimmerman 1995, p. 60: 'if expressions like "the water in Heraclitus's tub" are to find a foothold in the real world, they must be anchored either in particular physical objects, or else in particular sets or pluralities of physical objects. A theory of masses must therefore restrict itself to one or the other of these alternatives, or to some hybrid view that takes some occurrences of the forms "the K" or "$Sm\,K$" to denote sums and others to denote sets or to be plural referring terms.' To my knowledge, Zimmerman does not mention a hybrid theory of masses elsewhere in any of his published works.

18. I use 'aggregate' here as defined by the 'bonding interpretation': material substances a, b and c compose an aggregate d if and only if a, b and c are in contact with – or, perhaps, are bonded in a certain way to – one another. See Chapters 4 and 6 for detailed discussions of Aquinas on aggregates.

19. Zimmerman 1995, p. 91.

20. Of course, I am assuming that the substance theory of masses fares as well as the sum theory of masses where matters unrelated to what I have been discussing here are concerned.

SELECT BIBLIOGRAPHY

Abel, D. (1996) 'Intellectual substance as form of the body in Aquinas' *Proceedings of the American Catholic Philosophical Association* 69: 227–36.

Anderson, J. (trans.) (1975) *Summa contra gentiles*. Book Two: *Creation*, by Thomas Aquinas, Notre Dame, IN: University of Notre Dame Press.

Armstrong, D. (1978) *Nominalism and Realism* Cambridge: Cambridge University Press.

Baker, L. R. (1997) 'Why constitution is not identity' *The Journal of Philosophy* 94: 599–621.

—— (1999a) 'Unity without identity: a new look at material constitution' *Midwest Studies in Philosophy* 23: 144–65.

—— (2000) *Persons and Bodies: A Constitution View* Cambridge: Cambridge University Press.

—— (2002a) 'Précis of *Persons and Bodies: A Constitution View*' *Philosophy and Phenomenological Research* 64: 592–98.

—— (2002b) 'Replies' *Philosophy and Phenomenological Research* 64: 623–35.

Baldner, S. (1999) 'St Albert the Great and St Thomas Aquinas on the presence of elements in compounds' *Sapientia* 54: 41–57.

Bazan, B. C. (1997) 'The human soul: form and substance? Thomas Aquinas's critique of eclectic Aristotelianism' *Archives d'Histoire Doctrinale et Littéraire du Moyen Age* 64: 95–126.

Bobik, J. (1954) 'Dimensions in the individuation of bodily substances' *Philosophical Studies* 4: 60–79.

—— (1963) 'Matter and individuation', in E. McMullin (ed.), *The Concept of Matter*. Notre Dame, IN: University of Notre Dame Press, pp. 281–92.

—— (1965) *Aquinas on Being and Essence: A Translation and Interpretation*. Notre Dame, IND: University of Notre Dame Press.

—— (1998) *Aquinas on Matter and Form and the Elements* Notre Dame, IN: University of Notre Dame Press.

Brown, C. (2001) 'Aquinas on the individuation of non-living substances' *Proceedings of the American Catholic Philosophical Association* 75: 237–54.

Burge, T. (1972) 'Truth and mass terms' *Journal of Philosophy* 69: 263–82.

Burke, M. (1980) 'Cohabitation, stuff, and intermittent existence' *Mind* 89: 391–405.

—— (1992) 'Copper statues and pieces of copper: a challenge to the standard account' *Analysis* 52: 12–17.

—— (1994a) 'Preserving the principle of one object to a place: a novel account of the relations among objects, sorts, sortals, and persistence conditions' *Philosophy and Phenomenological Research* 54: 591–624 (reprinted in Rea 1997b).

—— (1994b) 'Dion and Theon: an essentialist solution to an ancient puzzle' *Journal of Philosophy* 91: 129–39.

—— (1997) 'Coinciding objects: reply to Lowe and Denkel' *Analysis* 57: 11–18.

Callus, D. (1961) 'The origins of the problem of the unity of form' *The Thomist* 24: 257–85.

Chisholm, R. (1998) 'Identity through time' in P. van Inwagen and D. Zimmerman (eds) *Metaphysics: The Big Questions* Malden: Blackwell, pp. 173–85.

Churchland, P. (1986) *Neurophilosophy: Toward a Unified Science of the Mind/Brain* Cambridge, MA: The MIT Press.

Clarke, W. (2001) *The One and the Many: A Contemporary Thomistic Metaphysics* Notre Dame, IN: University of Notre Dame Press.

Copleston, F. (1955) *Aquinas* London: Penguin Books.

Davies, B. (1992) *The Thought of Thomas Aquinas* Oxford: Clarendon Press.

Decaen, C. (2000) 'Elemental virtual presence in St Thomas' *The Thomist* 64: 271–300.

Deferrari, R. (1986) *A Latin–English Dictionary of St Thomas Aquinas* Boston, MA: St Paul Editions.

Deferrari, R. and Barry, M. (1948) *A Lexicon of St Thomas Aquinas* Baltimore: Lucas Printing.

Descartes, R. (1996) *Meditations on First Philosophy* trans. J. Cottingham Cambridge: Cambridge University Press.

Dewan, L. (1999) 'The individual as a mode of being according to Thomas Aquinas' *The Thomist* 63: 403–24.

Doepke, F. (1997) 'Spatially coinciding objects' in M. Rea (ed.), *Material Constitution: A Reader* Lanham: Rowman & Littlefield, pp. 10–24.

Edwards, S. (1977) 'Some medieval views on identity' *The New Scholasticism* 51: 62–74.

Geach, P. (1967–8) 'Identity' *Review of Metaphysics* 21: 3–12.

—— (1997) 'Reference and generality: an examination of some medieval and modern theories (selections)' in M. Rea (ed.) *Material Constitution: A Reader* Lanham: Rowman & Littlefield, pp. 305–12.

Gibbard, A. (1975) 'Contingent identity' *Journal of Philosophical Logic* 4: 187–221 (reprinted in Rea 1997b).

Gilson, E. (1994) *The Christian Philosophy of St Thomas Aquinas* 5th edn trans. L. Shook Notre Dame, IND: University of Notre Dame Press.

Gracia, J. (1988) *Individuality: An Essay on the Foundations of Metaphysics* Albany, NY: SUNY Press.

Hasker, W. (1999) *The Emergent Self* Ithaca, NY: Cornell University Press.

Heller, M. (1984) 'Temporal parts of four-dimensional objects' *Philosophical Studies* 46: 323–34 (reprinted in Rea 1997b).

Hirsch, E. (1995) 'Identity' in J. Kim and E. Sosa (eds) *A Companion to Metaphysics* Oxford: Blackwell, pp. 229–34.

—— (1999) 'Identity in the Talmud' *Midwest-Studies-in-Philosophy* 23: 166–80.

Hobbes, T. (1839) *The English Works of Thomas Hobbes*, vol. 1, *Concerning Body* ed. William Molesworth, London: Bohn.

Hoffman, J. and Rosenkrantz, G. (1997) *Substance: Its Nature and Existence* London: Routledge.

Hughes, C. (1996) 'Matter and individuation in Aquinas' *History-of-Philosophy Quarterly* 13: 1–16.

—— (1998) 'Matter and actuality in Aquinas' *Revue Internationale de Philosophie* 52: 269–86.

Klubertanz, G. (1963) *Introduction to the Philosophy of Being* New York: Appleton Century-Crofts.

Kretzmann, N. (1999) *The Metaphysics of Creation: Aquinas's Natural Theology in* Summa contra gentiles *II*, Oxford: Clarendon Press.

Kretzmann, N. and Stump, E. (eds) (1993) *The Cambridge Companion to Aquinas* Cambridge: Cambridge University Press.

Leclerc, I. (1969) 'The problem of the physical existent' *International Philosophical Quarterly* 9: 40–62.

Lewis, D. (1986) *On the Plurality of Worlds* Oxford: Blackwell.

—— (1998) 'The problem of temporary intrinsics: an excerpt from *On the Plurality of Worlds*' in P. van Inwagen and D. Zimmerman (eds) *Metaphysics: The Big Questions* Malden: Blackwell, pp. 204–5.

Lowe, E. (1998) *The Possibility of Metaphysics: Substance, Identity, and Time* Oxford: Clarendon Press.

McGovern, M. (1987) 'Prime matter in Aquinas' *Proceedings of the American Catholic Philosophical Association* 61: 221–34.

McInerny, R. (1996) *Aquinas and Analogy* Washington DC: Catholic University of America Press.

Matthews, G. (1982) 'Accidental unities' in M. Schofield and M. Nussbaum (eds) *Language and Logos* Cambridge: Cambridge University Press, pp. 251–62.

Myro, G. (1997) 'Identity and time' in M. Rea (ed.) *Material Constitution: A Reader* Lanham: Rowman & Littlefield, pp. 148–72.

O'Connor, T. (1994) 'Emergent properties' *American Philosophical Quarterly* 21: 91–104.

Olson, E. (1997) *The Human Animal* Oxford: Oxford University Press.

Pasnau, R. (2002) *Thomas Aquinas on Human Nature* Cambridge: Cambridge University Press.

Pegis, A. (1983) *St Thomas and the Problem of the Soul in the Thirteenth Century* Toronto: Pontifical Institute of Medieval Studies.

Plantinga, A. (1974) *The Nature of Necessity* Oxford: Clarendon Press.

Plutarch (1967) *Plutarch's Lives*, vol. 1. Cambridge, MA: Harvard University Press.

Rea, M. (1997a) 'Introduction' in M. Rea (ed.) *Material Constitution: A Reader* Lanham: Rowman & Littlefield, pp. xv–lvii.

—— ed. (1997b) *Material Constitution: A Reader* Lanham: Rowman & Littlefield.

—— (1998) 'Sameness without identity: an Aristotelian solution to the problem of material constitution' *Ratio* 11: 316–28.

—— (2000) 'Constitution and kind membership' *Philosophical Studies* 97: 169–93.

—— (2002) 'Lynne Baker on material constitution' *Philosophy and Phenomenological Research* 64: 607–14.

Reith, H. (1958) *The Metaphysics of St Thomas Aquinas* Milwaukee, WI: Bruce Publishing.

Renard, H. (1943) *The Philosophy of Being* Milwaukee, WI: Bruce Publishing.

Searle, J. (1992) *The Rediscovery of the Mind* Cambridge, MA: The MIT Press.

—— (1997) *The Mystery of Consciousness* New York: The New York Review of Books.

Sedley, D. (1982) 'The Stoic criterion of identity' *Phronesis* 27: 255–75.

Shoemaker, S. (1998) 'Personal identity: a materialist account' in P. van Inwagen and D. Zimmerman (eds) *Metaphysics: The Big Questions* Malden: Blackwell, pp. 296–309.

Sider, T. (2002) 'Review of *Persons and Bodies: A Constitution View*' by Lynne Rudder Baker *The Journal of Philosophy* 99: 45–8.

—— (2003) *Four-dimensionalism: An Ontology of Persistence and Time* Oxford: Oxford University Press.

Simons, P. (1987) *Parts: A Study in Ontology* Oxford: Clarendon Press.

Stump, E. (1989) *Dialectic and Its Place in the Development of Medieval Logic* Ithaca, NY: Cornell University Press.

—— (1995) 'Non-Cartesian substance dualism and materialism without reductionism' *Faith and Philosophy* 12: 505–31.

—— (2003) *Aquinas*. Arguments of the Philosophers, London: Routledge.

Stump, E. and Kretzmann, N. (1985) 'Absolute simplicity' *Faith and Philosophy* 2: 353–82.

Thomas Aquinas (1929–47) *Scriptum super libros Sententiarum* ed. P. Mandonnet and M. Moos, Paris: Lethielleux.

—— (1932) *De ente et essentia* Turin: Marietti.

—— (1934) *Summa contra gentiles* (Leonine edn) Rome: Desclee and Herder.

—— (1948–50) *Summa theologiae* ed. P. Carmello, Turin: Marietti.

—— (1948a) *Commentum in quatuor libros Sententiarum magistri Petri Lomdardi* (Parmae edn), vol. 2, New York: Musurgia Publishers.

—— (1948b) *In Aristotelis librum De Anima commentarium* ed. A. M. Pierotta, Turin: Marietti.

—— (1948c) *Summa theologica* trans. by the Fathers of the English Dominican Province Allen, TX: Christian Classics.

—— (1949a) *De principii naturae* ed. R. Perrier, Paris: Lethielleux.

—— (1949b) *Quaestio disputata de anima* in R. Spiazzi and others (eds) *Quaestiones disputatae*, vol. 2, Turin and Rome: Marietti.

—— (1949c) *Quaestio disputata de unione verbi incarnate* in R. Spiazzi and others (eds) *Quaestiones disputatae*, vol. 2, Turin and Rome: Marietti.

—— (1949d) *Quaestiones disputatae de potentia* in R. Spiazzi and others (eds) *Quaestiones disputatae*, vol. 2, Turin and Rome: Marietti.

—— (1949e) *Quaestio disputata de spiritualibus creaturis* in R. Spiazzi and others (eds) *Quaestiones disputatae*, vol. 2, Turin and Rome: Marietti.

—— (1949f) *On Spiritual Creatures* trans. M. Fitzpatrick and J. Wellmuth, Milwaukee, WI: Marquette University Press.

—— (1950a) *In duodecim libros Metaphysicorum Aristotelis expositio* ed. R. Cathala and R. M. Spiazzi, Turin: Marietti.

—— (1950b) *Compendium theologiae* in *Opera Omnia* (Parma edn), vol. 16, New York: Musurgia Publishers.

—— (1952a) *In Aristotelis libros Sententia super libros De caelo et mundo De generatione et corruptione Meteorologicorum* ed. R. Spiazzi, Turin: Marietti.

—— (1952b) *On the Power of God* trans. by the English Dominican Fathers. Westminster, MA: The Newman Press.

—— (1953) *In octo libros de physico auditu sive Physicorum Aristotelis commentaria*, ed. P. Angeli and M. Pirotta, Neapoli: M. D'Auria.

—— (1954a) *De mixtione elementorum* in R. Spiazzi (ed.), *Opscula philosophica* Turin: Marietti.

—— (1954b) *Responsio ad magistrum Joannem de Vercellis de 108 articulis*, in R. A. Verardo and R. M. Spiazzi (eds) *Opscula theologica* Turin: Marietti.

—— (1955) *Expositio super librum Boethii De Trinitate* ed. Bruno Decker, Leiden: E. J. Brill.

—— (1956) *Quaestiones quodlibetales* ed. R. Spiazzi, Turin: Marietti.

—— (1963) *Commentary on Aristotle's Physics* trans. R. Blackwell, R. Spath and W. Thirlkel, New Haven, CT: Yale University Press.

—— (1965) *Aquinas on Being and Essence: A Translation and Interpretation* ed. J. Bobik Notre Dame, IN: University of Notre Dame Press.

—— (1968) *On Being and Essence* trans. A. Maurer, Toronto: Pontifical Institute of Medieval Studies.

—— (1975a) *Summa contra gentiles*. Book One: *God* trans. A. Pegis, Notre Dame, IN: University of Notre Dame Press.

—— (1975b) *Summa contra gentiles*. Book Two: *Creation* trans. J. Anderson, Notre Dame, IN: University of Notre Dame Press.

—— (1975c) *Summa contra gentiles*. Book Four: *Salvation* trans. C. O'Neil, Notre Dame, IN: University of Notre Dame Press.

—— (1984) *Questions on the Soul* trans. J. Robb, Milwaukee, WI: Marquette University Press.

—— (1993) *Light of Faith: The Compendium of Theology* Manchester, NH: Sophia Institute Press.

—— (1994) *Commentary on Aristotle's De anima* trans. K. Foster and S. Humphries, Notre Dame, IND: Dumb Ox Books.

—— (1995) *Commentary on Aristotle's Metaphysics* trans. J. Rowan, Notre Dame, IN: Dumb Ox Books.

—— (1997) *Aquinas on Creation: Writings on the 'Sentences' of Peter Lombard* trans. S. Baldner and W. Carroll, Toronto: Pontifical Institute of Medieval Studies.

—— (1998a) *Principiis naturae* in J. Bobik (ed.) *Aquinas on Matter and Form and the Elements* Notre Dame, IN: University of Notre Dame Press.

—— (1998b) *De mixtionis elementorum* in J. Bobik (ed.) *Aquinas on Matter and Form and the Elements* Notre Dame, IN: University of Notre Dame Press.

—— (1999) *A Commentary on Aristotle's De anima* trans. R. Pasnau, New Haven, CT: Yale University Press.

Thomson, J. J. (1983) 'Parthood and identity across time' *Journal of Philosophy* 80: 201–20 (reprinted in Rea 1997b).

Torrell, J. (1996) *Saint Thomas Aquinas*, Vol. 1: *The Person and His Work* trans. R. Royal Washington DC: The Catholic University of America Press.

Van Cleve, J. (1995) 'Essence/accident' in J. Kim and E. Sosa (eds) *A Companion to Metaphysics* Oxford: Blackwell, pp. 136–8.

Van Inwagen, P. (1981) 'The doctrine of arbitrary undetached parts' *Pacific Philosophical Quarterly* 62: 123–37 (reprinted in Rea 1997b and Van Inwagen 2001).

—— (1990a) *Material Beings* Ithaca, NY: Cornell University Press.

—— (1990b) 'Four-dimensional objects' *Nous* 24: 245–55 (reprinted in Van Inwagen 2001).

—— (1993) *Metaphysics* Boulder, CO: Westview Press.

—— (1997a) 'Foreword' in M. Rea (ed.), *Material Constitution: A Reader*. Lanham: Rowman & Littlefield, pp. ix–xii.

—— (1998) *The Possibility of Resurrection and Other Essays in Christian Apologetics* Boulder, CO: Westview Press.

—— (2001) *Ontology, Identity, and Modality: Essays in Metaphysics* Cambridge: Cambridge University Press.

Veatch, H. B. (1974) 'Essentialism and the problem of individuation' *Proceedings of the American Catholic Philosophical Association* 48: 64–73.

White, K. (1995) 'Individuation in Aquinas's *Super Boetium De Trinitate*, q. 4' *American Catholic Philosophical Quarterly* 69: 543–56.

Wiggins, D. (1968) 'On being in the same place at the same time' *The Philosophical Review* 77: 90–5 (reprinted in Rea 1997b).

Wippel, J. (2000) *The Metaphysical Thought of Thomas Aquinas: From Finite Being to Uncreated Being* Washington DC: The Catholic University of America Press.

Yablo, S. (1998) 'Essentialism' in E. Craig (ed.), *Routledge Encyclopedia of Philosophy*, vol. 3, London: Routledge, pp. 417–22.

Zimmerman, D. W. (1995) 'Theories of masses and problems of constitution' *Philosophical Review* 104: 53–110.

—— (1996) 'Could extended objects be made out of simple parts? An argument for "atomless gunk"' *Philosophy and Phenomenological Research* 56: 1–29.

—— (1997) 'Coincident objects: could a stuff ontology help?' *Analysis* 57: 19–27.

—— (1998) 'Temporary intrinsics and presentism' in P. van Inwagen and D. Zimmerman (eds) *Metaphysics: The Big Questions* Malden: Blackwell, pp. 206–19.

—— (2002) '*Persons and Bodies*: constitution without mereology?' *Philosophy and Phenomenological Research* 64: 599–606.

INDEX